W9-CDF-662

the organic gardener

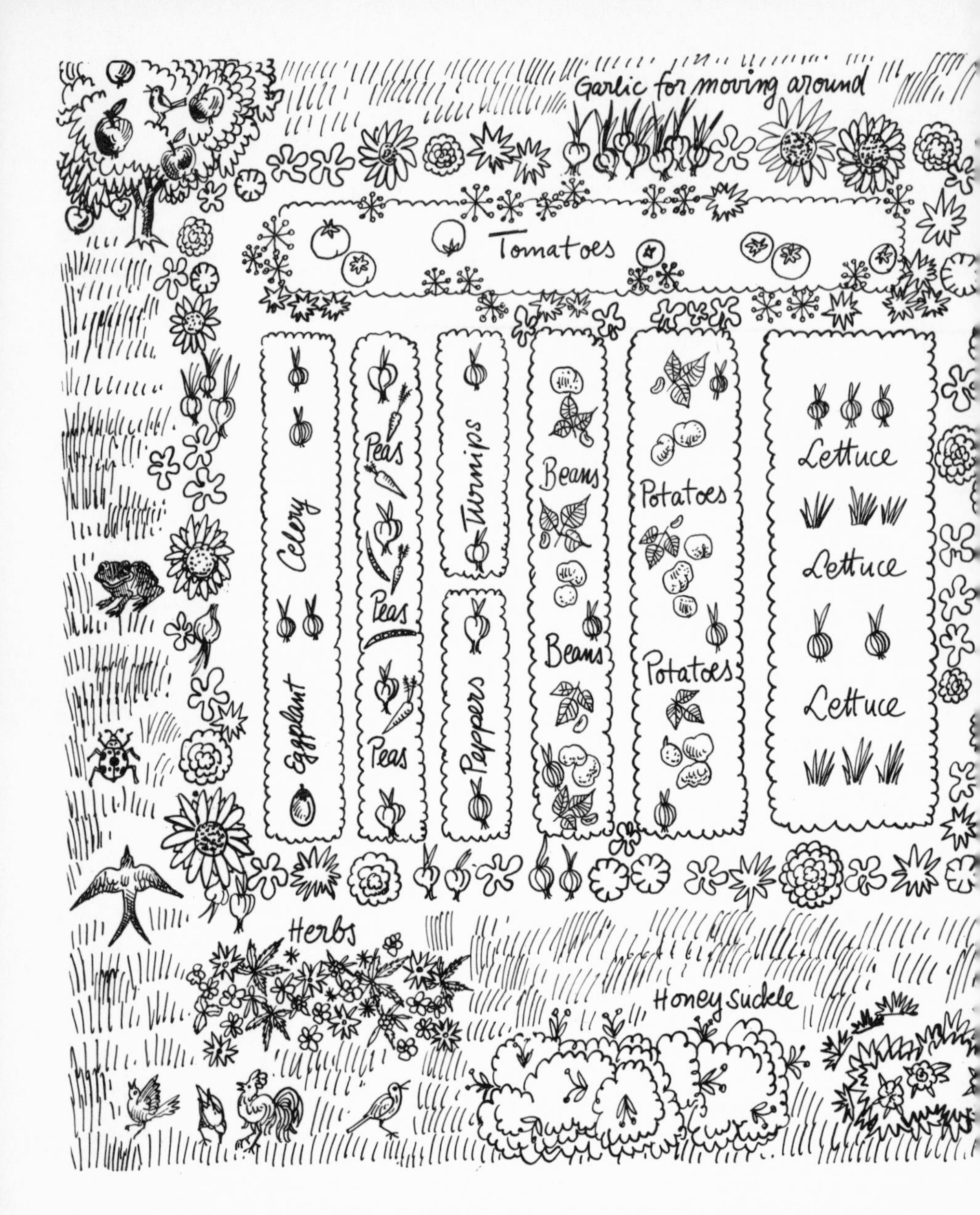

Alfred A. Knopf · New York · 1972

Drawings by Karl W. Stuecklen

the organic gardener

Catharine Osgood Foster

This is a Borzoi Book
published by
Alfred A. Knopf, Inc.

Published in the United States by
Alfred A. Knopf, Inc., New York,
and simultaneously in Canada
by Random House of Canada Limited, Toronto.
Distributed by Random House, Inc., New York.

Library of Congress Cataloging in Publication Data
Foster, Catharine Osgood, 1907–
The organic gardener.
Bibliography: p.
1. Organiculture. I. Title.
S605.5.F67 1972b 631.5'8 71-171142
ISBN 0-394-47210-1

Manufactured in the United States of America
Published May 31, 1972
Second Printing, October 1972

Grateful acknowledgment is made to the Rodale Press, Inc.
for permission to quote from J. I. Rodale's *The Organic Front*,
copyright © 1949 by J. I. Rodale.

For
Barbara,
Jill,
Linda,
Bob,
Fred, &
Tom—
good helpers,
teachers, &
gardeners

contents

illustrations

the organic gardener

chapter one

rich soil, few pests, and a varied environment

One day I woke up to the fact that I was an organic gardener without knowing it. I had been for years; for my husband and I had been farming and gardening according to old Vermont principles of respect for the land and the world of nature. We did things in ways that meant returning what we took away from the soil, and refraining from upsetting the balance. In the toolhouse we had some Bordeaux mixture and Black Leaf 40 for pest deterrent, but we never used them except very occasionally for some enormous invasion—usually on an exotic plant that didn't belong in Vermont, anyway. Though we did not systematically make compost piles, we did put all our organic residues on the land, because that was our frugal Yankee habit. Organic gardeners are now defined as those who use similar methods and who use neither poisons to kill pests and weeds nor premixed chemical fertilizers. If you are an organic gardener, you rely on the natural cycles of control and the natural fertilizers that come from the decomposition and mineralization of organic materials in the soil.

Our farm at that time was varied—with an orchard here, hen yards there, a big barn, several gardens, berry patches, hedges, a running brook, a swamp, and fields beyond stretching to pastures on the lower slopes of a wooded Vermont mountain. There was no monoculture anywhere. But we did have a lot of chickens. With variety like that there is an entire population of nature's controls alive and working all the time in the air, around the plants, in the soil, and in the healthy interchange of nutrients and antibiotics, in solution in the water of the soil. The soil structure and the soil solution, I know now, were continually replenished by the organic matter put out on the ground from the henhouses, the vegetable garbage, the mown grass, the autumn leaves, and the weeds we threw out. We had plenty of wasps, bees, ladybugs, a few praying mantises, and many

rich soil, few pests . . .

birds who ate lots of insects and weed seeds. Whenever the hens were let out at the proper season, they may have gobbled up a hundred asparagus beetles in an hour. We had no Japanese beetles then. And ducks and geese and goats were the only weed-killers we knew about.

The gardens were on slopes, well drained and well aired. They were well-nourished gardens, too, especially the one we put on the spot where we tore down an old henhouse. The organic matter there, after a winter under a thick blanket of snow, was totally composted by the time we got ready to plant. Organic gardeners who compost systematically have heaps and a routine of composting. In those days we just put manure out as we had it. And we wouldn't have thought of having a garbage man, for we trusted the land to take care of the wastes we put on it. It did.

Besides, we were busy. There was teaching to do, articles to write, research or lessons to work at, the hens to take care of, and birds to watch—protecting the birds, feeding them and enticing them in with places to nest, bird feeders, and wild berries to eat, which they actually prefer to the less tangy cultivated cherries, blueberries, or raspberries. In fact, protection of the birds was one main reason for our refusal to use poisons. Once in a while I'd get an urge to spray, but my husband wouldn't permit it. He'd remind me that we wanted to protect our dogs, and the other animals, and the birds; and we wanted clean fruits, chickens, and vegetables for our own use and for the neighbors who were our customers.

We had never heard of aldrin, dieldrin, DDT, or other hard chlorinated hydrocarbon pesticides. When they came on the market, I believed the ads and bought some sort of mixture with DDT in it, but my husband forbade that, too. Ever since, the magazines and journals and a good many of the books that have come into our house have revealed with increasing alarm what these poisons are made of, what they do, how they build up instead of being biodegradable, and how they affect biological systems and threaten the entire food chain from bacteria to eagles. We saw one article reporting that in a certain New York City restaurant a few years ago there was not one item on the menu except the black coffee which did not have DDT in it, according to lab tests done on all foods served there.

I began to learn new ways to do things because I had to. I read a great deal about organic gardening, visited gardeners who used organic in preference to chemical methods, and saw that many of them, too, learned new ways because they had to—and wanted to. Many remembered, as I did, the dust bowl in the Middle West, and deplored the loss and exploitation of good topsoil in many, many parts of this country. Many were reading, as we were, about the dire, unforeseen side effects of the hard pesticides.

. . . and a varied environment

Some, like neighbors of ours on a gravelly, shaly side hill, were challenged by the problems of building a whole new soil. They discovered that manure, compost, and the newly made soil which they created themselves was considerably more satisfactory and cheaper than buying tons of topsoil to be brought in. The previous owners of their place had bought some topsoil for the lawn, but the area where the new family wanted to put their vegetable garden was so gravelly it was impossible. The minute you start scavenging for a compost heap you become expert at finding sources for the organic matter you want to use for it. These gardeners found a goat farm, the dump where the town's street department put autumn leaves, those neighbors who did not want their lawn clippings, and a chicken farmer who would let them have all the litter and manure they'd take away if they'd gather it and pile it in the truck themselves. Like everyone else, they had plenty of weeds to add to the pile and in the fall some old mulch from their first little garden. I watched all this process with great interest, and when the wife, who is a member of our garden club, talked on "Conservation, Backyard Style," I learned a lot of new tricks. Now, after several years, that garden in the backyard looks fine.

When we moved to a new place, we were glad to find that the previous owners had been careful gardeners who built up an excellent soil. In fact it was so good that Japanese beetles had moved right in. They were getting pesty on the asparagus and grapes, but that year our county agent asked members of the garden club to try out the newly discovered milky spore disease which parasitizes the larvae of the beetle and puts them out of commission. It has nothing to do with milk; it just makes the larvae look white.

This was my first experience with a biological control that was called by that name, though birds and wasps, ladybugs and ants were always helpers in maintaining populations that would get out of hand completely if not preyed upon. And the beneficial bacteria and fungi in the soil are certainly biological controls all the time, whatever you call them. So are garlic and the curious antibiotic that is dis-

rich soil, few pests . . .

charged from it, and so is the tansy that fends off the pesty insects called aphids. So are what are called trap plants, like the dill which attracts tomato worms away from tomatoes, and the white roses which, if you can bear to think of its happening, will distract your Japanese beetles away from your grapes and corn and asparagus and zinnias. Well, so are the zinnias. Even more distracting is the knotweed that will come up among your zinnias if you let it.

Maybe it was luck or maybe it was the right day for me, but at just the right time I heard a lecture which introduced me to organic sprays made of garlic, onion, and red pepper, and even plain soap and water. The aphids had come, so I tried these sprays. They worked—the soapy water on the lupines, and plain water followed by garlic and onion on the string beans. (See p. 70 for recipes.)

The next year I began taking classes in plant physiology to find out what was going on and what the processes were which were affected or distorted by good or bad practices in the garden and around the yard.

Now, if trouble strikes, I don't have to reach for the bottle or aerosol can to cure the ill with medicine, or rather poison. Nor do I dream of telephoning for an airplane to spray the entire place or of ordering one of those little tractors shown in ads with pretty models who are supposed to be riding up and down their yards spraying everything in sight. I go out and spread wood ashes, or transplant nasturtiums or marigolds to the trouble spot, or get out the hose and give infested plants a good washdown. My husband likes to think up ways of outfoxing the woodchucks and rabbits—by sprinkling blood meal around their holes or down by a row of lettuce—or of trapping insects like earwigs in dark lengths of rhubarb stem, and fending off snails and slugs with a three-inch barrier of sharp salty sand, which they'll refuse to come through with their soft, tender bodies. He will not get out the gun, and he won't spray poison or poison the corn seeds. The crows he keeps tame and well fed, and they come to one of our feeders on a table under a pear tree six at a time in the middle of summer. When the cardinals begin to whistle at feeding time, the crows soon tune up way down in the woods, and come in after we whistle, too.

Today I also know I was a natural foods cook—or nearly one—though I didn't know that, either. From some home economics class I took in school, or from some article

. . . and a varied environment

in one of the farmers' magazines, I'd learned that you'd throw nutritious vitamins and minerals down the drain if you strained your vegetables without saving the water they were cooked in. I didn't always use this liquid for soups and gravies, so I invented other ways to use it. You can put it in biscuits and muffins as part of your liquid ingredient. You can save it for cooking the next vegetable, if the tastes don't swear at each other. If it won't be used for a day or so, put it in an ice tray and freeze it. In the summer, when you blanch vegetables for freezing, you can use the water over and over, and afterward use some of it to thin mayonnaise, white sauce, or cottage cheese. And what's left, use for blending a green drink of herbs and greens the way Euell Gibbons makes them when he is stalking the wild asparagus or the healthful herbs. (See Appendix.) I have learned to do some of these in recent years, but in the days when I didn't get home from work till nearly six o'clock, I mostly used the pressure cooker which takes hardly any water, anyhow. I learned not to put salt in until after the cooking. If you stand over the pot and watch it, you can boil that little bit of water away at the last minute and have no problem at all about what to do with it. Gradually you also learn to use for organic purposes all those nutrients that used to be wastes.

Soften the dogs' dry kibbles with the extra cooking liquid. Put it in the pail of stuff that goes to the chickens or goats if you keep them. Put it on the compost heap, or right on the garden. Use it for making organic sprays, herb teas (if mild enough in taste), and to water your house plants. You might even dribble it out of the watering can onto the leaves, in the hope that its nutrients would make a foliar feeding spray. You could run an experiment to find out. (See Chapter 2.)

The cooking I did was simple, somewhat lazy and parsimonious, and reflected the fact that my edition of Fanny Farmer is very ancient. I cooked with what happened to be around, and a lot of it came from the garden, hedgerows, or henhouse. We did not have, never have had, a TV, so I was never tempted to try out all the newfangled, adulterated, souped-up timesavers that other women have been inveigled to try. I did try a few I saw on the shelves at the market, but there was resistance again from my Yankee husband, so I went right back to the natural ways I'd learned before. Besides, we both like the taste of foods prepared

rich soil, few pests . . .

fresh, of whole grain flours, pure maple syrup, fresh fruits, fresh vegetables, fresh eggs. And never having had a TV, we've never had a TV dinner, either.

We had farmer friends up and down the road where we lived, and an expert old German nurseryman as a neighbor at the top of the hill who gave us suggestions about when to plant, when to harvest, what varieties were good for our zone, and good tricks for cultivation. These farmers all used fresh or home-canned foods and stuck to the old ways of doing things, too. What I know now is that their gentle conservative old ways were, for the most part, organic gardening ways, so I believe that it is perfectly natural and easy for anyone wanting to grow his own vegetables to use these traditional methods. In those days many of us fell for superphosphate as a fertilizer because it worked fast and we could see results. Organic gardeners of today, preferring natural, untreated materials, use rock phosphate instead. It works all right, but more slowly, and it avoids an unhelpful residue.

As I see it, the only real difference between a pure organic gardener—a disciple of Sir Albert Howard of Indore, India; or Ehrenfried Pfeiffer, of the Bio-Dynamic gardening group; or the Rodales of the Organic Gardening and Farming Press in Emmaus, Pennsylvania; or any of the other old-time founders of Friends of the Land and the organic gardening movement—and a plain old conservative Yankee farmer is just this: that the organic gardener has a philosophical and scientific conviction against using chemical fertilizers and pesticidal poisons while the Yankee has doubts and suspicions, a deep satisfaction and certainty about what he has already done with success, and a respect for the wholeness of nature which he never verbalizes as the disciples do.

As long as the old-timer was feeding himself and his family, his stock and his chickens, he stuck to the good old ways, just as the organic farmer did. At our farm we knew that minerals in the heap of scraps to go out to the chickens were good for the chickens, and for the soil, too, when they again became available in manure to dress the land. We also knew, with or without a detailed knowledge of why, that it is good for the soil to manure it. We also knew that humus was good, whether or not we understood the structure of the soil and what makes for good nutrients for plants and for the animals and men who eat them. Whereas confirmed organic gardeners have carefully run compost heaps, farmers have dung heaps or whatever their state milk laws will permit.

. . . and a varied environment

After we moved from the farm to the rather suburban place where we live now, we did feel the pressures to switch to chemical gardening more acutely than ever before. People up and down our road were starting patios or putting in swimming pools and beginning to talk about getting rid of mosquitoes. The organic gardener wouldn't think of using any but herb sprays on areas where mosquito larvae breed. And now most garden club members and conservationists wouldn't either. We'd rather bring in young toads, and do; for they eat up mosquito larvae by the hundreds and even by the thousands. They have to be moved very young, for they don't like change after they grow up—a good thing, for once they are at your place, they do not want to leave. You can entice swallows by leaving your barn or garage door open, and watch them dart over mosquito areas and consume those insects in quantity. Dragonflies eat mosquitoes, and so do frogs and fish.

Many suburban people, and now many newly rural people, have only recently lived where nature and its rough ways are a daily encounter. When they go outdoors and find mosquitoes and other creatures sharing the environment they want for themselves, that puts them in a fighting mood. The organic gardener does try, I believe, to have a less belligerent way of dealing with things, and to try to fit in with the patterns that exist without man's interference. We stay outdoors more, at more times of day, and I've even advised complainers to stay indoors—or at least way away from where the mosquitoes are. And a lot of us are learning that nothing is perfect. I find it pleasant to know that the only thing perfect in nature is the total, overall design whose exquisite interlacing intricacy is beyond comprehension and beyond compare.

An artificial violet, a paper rose, or a plastic ear of corn can be changeless and spotless. Also ugly and utterly lifeless. Any live leaf or live string bean may have spots on it or bites taken out of it. I have heard a young housewife say *hurray* when she discovered that the carrots she was buying had some bumps and imperfections. I've said it myself, and hoped that the imperfections meant that here were some vegetables that had not been drenched with pesticides to make them smooth and unbitten.

We don't want our vegetables to look as smooth and flawless as plastic, and we wish the public relations men and chemical company salesmen would stop saying that we do. The more organic gardeners learn about gardening, the

rich soil, few pests . . .

more we realize that not only the use of pesticides but also the plant breeders' and truck-garden farmers' preference for vegetables that ship well or last long on the shelf are subjecting us to stringy cabbage, rubbery lettuce, huge peas, and all the other vegetables that are fibrous, tasteless, and tough. Now we grow our own.

I've talked to any number of people—and gardeners, too—who think they cannot have unsprayed, or rather unpoisoned, gardens without having something come and ruin them. There are lots of answers to this fear.

The first is: There are dozens of other preventives if you get a shortage of water, a crop that won't grow, or an insect invasion—from plain hosing down the plants to replanting or fertilizing to setting traps, including trap plants and companionate plantings of tomatoes and cucumbers, for example. You can also use black lights to catch egg-laying moths, gooey tanglefoot to trap caterpillars, or a little beer in an almost-empty can to entice snails and slugs to do themselves in.

Another point is: What if you do have to replant here and there or what if you do have a few bugs? A few bugs won't hurt you. Pick them off and drop them in a jar of water with a film of kerosene over it. Or let some of them eat up a few plants—the ones they choose will be weaklings, anyway, which you would pull out later. Unless you have a great big farm operation you won't get a huge invasion such as those that sometimes come to large fields of a single crop where the feasting is all too favorable.

One of the best answers of the organic gardener to objectors is that if you have a very good, rich, healthy soil, your plants will be rugged and healthy and unattractive to insect predators. It works. I've seen it over and over in many home gardens. In my own garden I've seen that the lowest, senescent leaves on the runtiest plants are the ones that get bitten and attacked first. Those worst leaves on the smallest of the squash vines or cucumber vines are the ones the cucumber beetles will go right for. Pick the predators off and pickle them—in that kerosene brine. If it is dry weather, give the plants a good watering, too. Or if it is pollination time, when plants always need lots of water.

The healthy soil and garden will be at the top of their form in a yard full of creatures and plants that live at your place with you. You do what you can to keep them that way, discouraging few of the living creatures who come, and allowing the patterns of coexistence to settle as many problems as possible. This is the organic attitude.

. . . and a varied environment

A last answer is: you know you don't want to mess up such marvelous patterns with poisons, and you know you don't want to interrupt the food chains of all the creatures in your yard or to eat lettuce yourself that has been spattered with some sort of DDT or other hard chlorinated hydrocarbon pesticide. If some of it washes into the ground, it may reach the bacteria, fungi, actinomycetes or little microorganisms of a kind between fungi and bacteria, as well as molds, earthworms, shrews, perhaps mice and moles, and all others who keep the soil in condition and the cycles in operation.

A few people, and very thoughtful many of them are, still are worried about whether the United States has to be or should be the breadbasket for the world and can continue to be in spite of recent horse plagues, cattle fevers, and corn blights. They worry whether or not American large-crop monoculture is inevitable for alleviating the plight of half the people of starving populations. Some of them believe that it cannot go on without pesticides (which they call economic poisons) and chemical fertilizers.

We are in a dilemma, they fear, and they overestimate the role the United States can or should play. Some aspects of the problem they sometimes neglect: for instance, the question of the water shortage looming ahead; the starvation in this country and the appalling malnutrition alongside our affluence; and the fact that we have mountains of compostable refuse of a hundred or so kinds, millions of gallons of high-nitrogen effluents, billions of tons of rock phosphate we could put to use.

We have passed laws (though they do not operate very well) to control activities of poisoning with pesticides and herbicides. We have found many new sources for other kinds of pest control. And as a happy note, 75 percent of the agricultural research at the Bethesda, Maryland, labs of the U. S. Department of Agriculture is now devoted to developing more and better biological controls for the economically important pests. Good progress is being made, I hear.

When I was asked to write this book, my first thought was that all the practical things that could be said about organic gardening had already been said, especially in the books and the magazine *Organic Gardening and Farming* published by the Rodale Press, and in guides by Beatrice Trum Hunter, Samuel Ogden, Euell Gibbons, and Alicia Bay

rich soil, few pests . . .

Laurel. Then I began thinking about all the young people I know who are boldly launching out into a new style of living, with gardening as one of their main occupations. There are hints that can help any beginner, and there are some cautions and reassurances to give. Also there are all the fascinating explanations for the various methods you use. When I made up my mind to do it, I found it very exciting to review suggestions from other experienced gardeners and to comb through books by plant physiologists and soil experts to find what they have said about the roles of organic matter in the soil and as taken up into the plant, and about the predation patterns among plants and animals which keep everything cycling.

The chapters that follow are addressed to several kinds of readers: to old-time gardeners looking for suggestions from some of the recent literature about biological controls; to enthusiastic but inexperienced young natural-foods gardeners looking for ways to get started and practical things to do; to those still undecided about the kind of gardener they want to be, who are looking for some better arguments than they have heard before for organic gardening without pesticides and for using a composting method in preference to a chemical-fertilizer one. It is natural to want valid reasons for saying you don't have to use poison sprays because there are other ways for man to help keep down pests; and valid reasons for believing that the nutrient qualities and structure of your soil are better if you add organic matter and compost to it. For the most part these chapters will be for those who want gardening and farming to move ahead from a fixation on chemicals and pesticides and encompass again the natural riches available everywhere—from the banana skin in your own garbage pail to the monumental and potentially recyclable wastes from our big cities.

After some suggestions about things to do to get ready to garden, there are chapters on composting or making your own nutrient mixtures, on the soil and what it does for us and for our gardens, and on the garden itself, its vegetables and other benefits.

With some know-how, enthusiasm, and the recognition that a few possible failures are not going to hurt things much if you stick to natural methods, you will find you can have in your garden what you want in it. I hope everything at your place will be blooming with good riches, delicate fresh foods, and a sense of well-being because all that you'll have there belongs there.

chapter two

getting started with indoor gardening and basic botany

Gardeners always learn that nature is not a plaything of man. Even though some still ride over their fields in big machines and fling out chemicals and pesticides onto the land, many of the rest of us have made up our minds to cooperate with as many natural forces as possible, and leave out the dubious chemicals. Doing organic gardening is a way of taking part in the cycles of nature's overall garden, we feel.

As organic gardeners we take positive steps to see that the capacities of nature are not overexploited and polluted, and to do this we need to learn as much as possible about what these capacities are. The practices of gardening show you what they are; a knowledge of the natural processes behind the practices will show you why organic gardeners choose to do what they do. The purpose of this book is to enlarge your acquaintance with both, and to help you over the humps, especially when you want to take things too seriously, or consult too many experts, or panic. I want to give you some hints, and some of the facts I've found that appear to be pretty good explanations about the soil and the growth habits of plants; most of all I want to make the enterprise of gardening rewarding, relaxing, and exciting.

start on any scale, any place, in any season

Start in as small a way as you like—on the kitchen counter, with a few radish seeds on a wet blotter or with plans for conversion of a suburban yard to new methods free of pesticides and dubious chemicals. You can set up a small greenhouse under fluorescent lights to start flowers and vegetables in organic Ferto-pots for later transplanting to the garden or window box or terrace planter. You can begin small-scale composting by saving your vegetable garbage and hunting around your neighborhood for humus-making materials like wood chips to scavenge. Such materials can be put in a plastic bag or in the beginnings of a compost bin you start to build out of wood or concrete blocks. You can study seed catalogs, draw up layouts and planting schedules for a small city yard, park-allotment plot, or big south-sloping field on a new homestead. However you begin, look forward to a lush, pestless, well-composted, and fertilized garden controlled without poisons or additives, organically nourished and blooming with clean, bright flowers and ripe delicious vegetables and fruits.

You can enter the yearly revolving cycle of nature's twelve-month garden in winter, spring, summer, or fall, and do things for the sake of a garden of your own that are appropriate to any of those times of year. If it is spring, you hurry to prepare a place, order seeds, and get right out to plant—and it is all rather hasty and may raise some difficulties if you haven't taken time to plan properly ahead of time. If it is summer, you can travel around and see from other people's gardens what you'd like to do the following year, watch what grows well on different parts of your land, make plans, start composting, and prepare for fall plowing and a cover crop, perhaps. You might also try some small quick crops like cress, beans, and radishes on cleared ground. If you start in the fall, composting might be your main effort, since it is the season for plentiful leaves to include in your pile, and again plan on fall plowing and a cover crop. In either of these seasons you can study your land for possible needs for drainage or irrigation, for the best locations of the future garden and fruit and nut trees, and perhaps try to start growing or transplanting materials you can harvest to make organic sprays for use the following year.

If you begin in winter, it is, in most areas of the country, a period of lull when there can be time to spend away from the outside chores that keep demanding attention in other seasons. But as a full-fledged organic gardener, you will find that summer chores are very much reduced—such as weeding and spraying—though you will still need to do some early-morning trap inspecting and bug gathering. Many of the suggestions I am going to make for the winter can, of course, be done in other seasons, from running experiments to learn what plants do under certain lights and heats to trying your hand at plant propagation.

All my suggestions are intended to get you started with experiments in which you can teach yourself about the principles of plant growth and the plant structures that make them function. You will remember what gardening practices are best because you will have done things for yourself and on a small enough scale so that every step can be clear in detail. You will have successes and failures, but no error or loss is tremendous—though you will discover that after you have got to know your plants, you will feel a sadness when something goes wrong with them. In winter indoors, you can usually make amends if something does go wrong, and you will have plenty of time to meditate about what is happening, and what can happen. You will always be learning how much more there is to learn.

starting . . .

plant radishes on a blotter

To learn about seeds, how they take up water and open up to grow, plant some radishes. All they need is water at first. Put them on a wet piece of towel or a wet blotter, or in a saucer with water; the blotter is best, for it controls the moisture. A cover of some sort is needed, either plastic or glass. A petri dish like those used in labs is good, because it is glass and you can watch what is happening. Keep the blotter moist, so the seeds will survive, germinate, sprout, and send out a root with root hairs. At this stage no nutrients need be added; all the nourishment the little plant needs is stored in the endosperm body of the seed. The radish, like many other plants, has two seed leaves, or two halves to the seed which separate when the embryo sends forth a root in one direction and then a stem in the other.

After the separation you can watch this fairly easily—especially if the seed coat has also slipped off. The root is oriented toward the earth and will grow down, if possible; the stem goes up, toward the light. On a horizontal petri dish the roots seem to go out and around. Once you get used to this radish and the way it grows, it might be a good idea to start some bean seeds. They are larger, and their structure is slightly different. When the seed coat has absorbed enough water to be soft and easily pushed off the core of the seed, gently open it and watch how the young plant growing from the embryo or germ gets going when it, too, has absorbed enough water to initiate the change. Once started, be sure not to let your seeds dry out. Try anything around the house—orange, lemon and grapefruit seeds, avocado, squash, wheat berries.

After radish seeds have been growing a little while, the root hairs are spectacular. Though none is more than a half-inch long, there are so many of them, and their surface exposure is so extensive, that each of these little plants is already on the way to its miles of surface exposure to the soil solution which would nourish it if it were in the ground. When in the ground, the root hairs form close bonds with the soil particles that yield nutrients to them. Each hair is a rounded protrusion of a single surface cell on the root. Most plant cells are squarish or oblong; but these root-hair cells have this extra protrusion. The wonder is that it is all part of one T-shaped cell and that each one is an efficient arrangement for intake of nutrients directly into the plant.

Both in the kitchen and outdoors in the garden the things you do to help your plants have to be scaled for entry into these tiny root-hair cells.

Petri Dish with Radishes
Showing Root Hair Cells

You will see, as the root grows, that its zone of root hairs changes—for they do not grow out from every part of a root. Old root-hair cells die and new ones are formed in the part of the root that is growing fastest, the zone of maturation, so-called, near each root tip. A rye plant studied by H. J. Dittmer had 14 million rootlets with a total length of 380 miles; he calculated that their root hairs ran to the billions, probably 14 or 15 billion.

The few hundred you may see on your radishes will be absorbing water and, later, nutrients for the plant. Since the roots are not buried you could move these whole little plants however you want without injuring the root hairs. You can finally plant them without doing much damage. But *planting radishes . . .*

if planted in the ground and then moved, they would suffer quite a lot from moving. Many of the tiny root hairs would get pulled off because in soil they will have forced their way up between the incredibly thin layers of clay into which nutrients are absorbed.

The little radish plants grow by cell elongation, getting the energy to do so from the stored carbohydrates in the endosperm of the seed. When the leaves form, if they are in the light, they turn green. This means that the chlorophyll-forming structures are now being created, and that photosynthesis, the making of the plant's own food, has begun. If given some complete plant nutrient, or fish emulsion in the water, the growth would go on.

Once in a while you have to expect failures. I've had zinnias growing in a petri dish develop brown rootlets and seem to lose their root hairs. Even so, in the moist atmosphere, the early first leaves kept going several days after the roots began to deteriorate. Though most will die, two or three can survive; the question is: How?

You begin to speculate when this sort of thing happens, and then you may plant over again, and compare what happens the next time. You can always learn from these failures. In the winter, on your experiment table, they are not serious. You just cross them off, put the ruined stuff on your compost heap, and start again.

now plant some seeds in potting soil

This is the right time for planting some seeds in a good potting soil, to see how they grow in that. Mix 7 parts loam, 3 parts peat, and 2 parts sand. For each 12 cups, add half a cup of compost or dried cow manure. All these you can now buy at garden centers and many chain stores. In order to watch the growth, use a plastic cocktail glass, with some holes knocked out in the bottom for aeration and drainage. Since the roots will shun the light, wrap up the container in foil to keep the soil dark. Then you can simply remove it whenever you want to look and see how the rootlets and root hairs are growing. I started several squash seeds this way last winter in good soil (but a small amount), and it was very interesting to watch how rapidly the whole container filled up with roots. This made the plants pot-bound. That is, they were crowded, or the three that survived were. The effect was good in that the competition for nutrients by what must have been yards of rootlets made the stems short, the plants small, and the blossoms come early. The one I transplanted to a slightly larger pot bloomed for two months. (This phenomenon of a plant's flowering when undernourished is often used by greenhouse gardeners to force blooming.) Eventually the stem got quite woody, probably at the time it began to develop a nitrogen deficiency, though the blooming itself would indicate a nitrogen deficiency already.

plant a potato

In a much larger clay pot I planted a potato in a rich, light potting soil. I didn't bother to cut the tuber in pieces; I just put it in whole and waited. In a few days two sprouts ap-

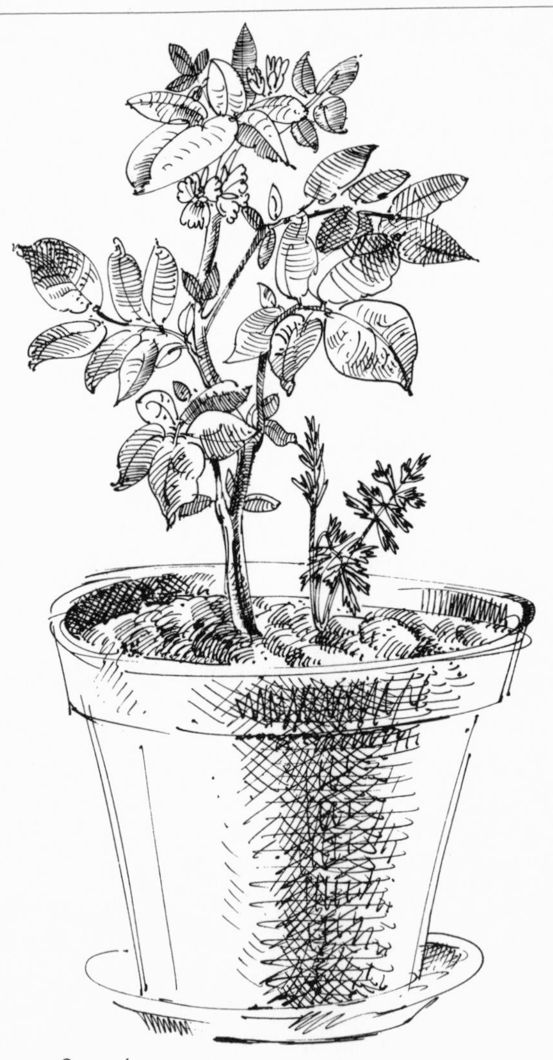

Potato and Carrot

planting potatoes . . .

peared. One was short, and never grew more than a couple of inches. The other shot up to a height of almost three feet in a matter of weeks. It blossomed once, and then the delicate white petals fell off and the plant stopped growing and just stayed there as a tall green accent in the kitchen window. This pot, too, filled up with a mat of roots. It required an enormous amount of water, and I kept the feathery-leaved top slice of a carrot pressed into the soil as an indicator. When there was enough water, the little green leaves that grew out of the carrot stood up and looked fine and healthy. When the pot got dry, down they drooped and looked about ready to die. I had waited for weeks for the top to die down as potato plants do in the garden when it nears the time to dig. By mid-June the leaves had begun to yellow and drop off. I pulled it up and saw nothing but a nubbin. What could you expect? Then Tom felt around in the soft earth and found two potatoes. One of the two was clean, scabless, and full-sized; the other was a very small one, also clean and unblemished and good. Next time I plan to grow potatoes in three different potting mixtures and compare them.

plant several tomatoes

Whether or not you plan to start your own tomato plants to set out in the garden later, start a few for experimental watching anyway. Try various ways to encourage them to flower early—on different plants. Give one plant a cold treatment after it reaches five inches in height. Keep it cool for two or three weeks. On another, water scantily, for four days, to suspend leaf growth temporarily. Take some of the leaves off another—especially the young expanding leaves. Then see whether these practices improve or inhibit the flowering in any way. "Disbudding" tomatoes has been known to increase the number of flowers, as well as to move up the flowering date from 80 to 50 days (especially if only two or three leaves were left on the plant). Then try giving a plant two or more days in one, by turning on fluorescent Vitalite lamps or Sylvania Gro-Lux during the night for an hour or putting the plant in a dark closet during the day for an hour, or both. All these experiments will not necessarily work, but they are interesting to try. A plant is a living thing, not a piece of machinery to tinker with. Such factors as intensity of light, degree of photosynthesis in the parts you leave on the plant, the action of various enzymes and pigments, and a hundred other factors you are not thinking of will be influencing the plant all the time, too. Last of all *planting tomatoes . . .*

try a sugar solution or apple juice, or even try putting a fresh, young, cut apple inside a plastic bag tied around a tomato plant. (If those won't induce flowering in a tomato, try the top of a pineapple. And if cold treatment won't induce flowering in a tomato, try it with peas.)

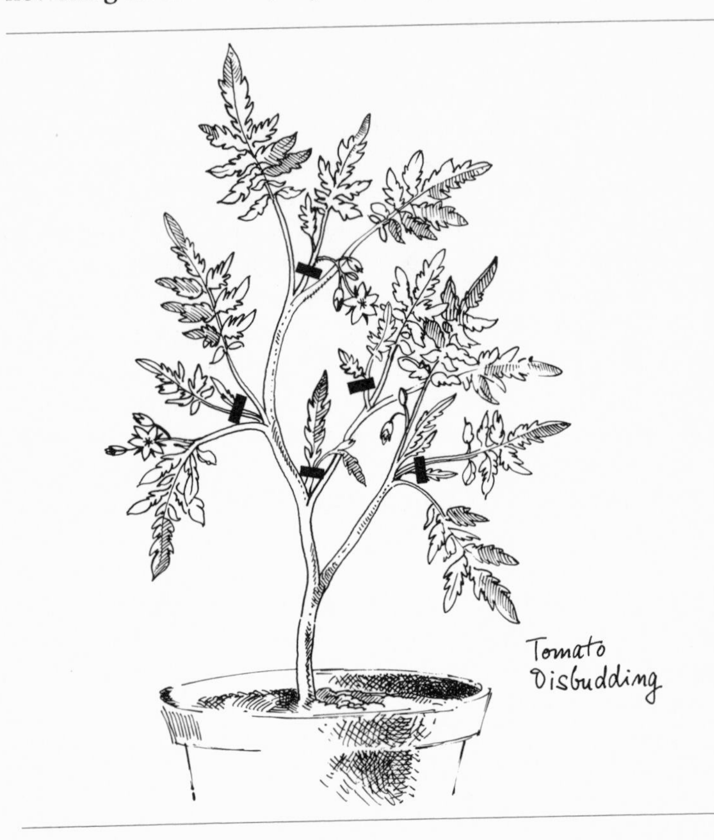

try various soil mixtures

For anyone just beginning to learn about organic plant growing, experiments with adding compost, dried manure, cottonseed meal, or other nutrients to pots of plants on the windowsill are well worth the effort. You can teach yourself what nitrogen, phosphorus, and other nutrients do, and compare leaf size and color, sturdiness of stem, and general well-being in pots treated differently. Making a new soil mixture and repotting your plants is also instructive. It is a good idea to keep a big pan of potting soil or a mixture of soil, leaf mold, and compost under a shelf or in some cupboard so you have it not only to add to when you make a scavenging discovery, but also to be handy when you need a new source of supply. Keep experimenting with different mixtures of soil and sand and with perlite, peat, vermiculite, and moss in different proportions to the basic loam. More will be said about what is in the soil and in its components in later chapters.

plant marigolds, mints, and tangy herbs

Aside from the experiments, you can begin plants that you will want for your cooking and for your garden later on. Marigolds are important as repellent plants and for use in organic sprays. Nasturtiums are also good to start for use outside later, and if they bloom, they brighten up your kitchen at the time of year you most need it. Those given a good rich soil and plenty of nutrients at frequent intervals will do well. Those planted in small pots, with a poverty of nutrients, will be stunted and eventually yellowish.

The herbs you can grow on your kitchen shelf for use in your cooking and for planting outside later are those that mind neither the heat when young nor the shock of transplanting. Those to try would include lemon balm, basil, borage, caraway, cress, dandelion, dill, hyssop, lovage, marjoram, parsley, rosemary, sorrel, French tarragon, and woodruff. Parsley and lovage may be moved to a shady spot in your yard, but most of the rest of them should be gradually moved into the direction of the sun. Do not try such delicate plants as chervil or bother with the hardy perennials available wild along so many of our roadsides, for example, blue-flowered chicory.

learning about light

After some of your plants are up, try moving them around to different exposures. A southern window is of course very useful for giving the plant plenty of sun, but since you can never quite achieve overhead sunlight, the growing plants will stretch out of shape in fulfilling their instinct to go toward the light. What happens is that a hormone responds to the need and acts on the cells on the side of the stem away from the light. These cells then elongate and thus enable the plant to bend and move in the direction it seeks to go. Keep turning the plants every day to compensate. Once the stems grow long, they stay long until the rest of the plant catches up—if it does. A plant needs direct, overhead light to slow down this stretching. If you can rig up a fluorescent light just over a shelf or your kitchen table where you can put your plants, try that too. The light that provides the full spectrum (including near-ultraviolet rays) is the best to use. Both plants and humans are believed to have better health *light . . .* under this full-spectrum light. Some people are having fairly good luck with full-spectrum Vitalite (used even in hospitals and offices now), Gro-Lux, and Plant-Gro. For imitation of full-spectrum daylight, you do not want plain fluorescent bulbs, and certainly not pink ones. Some of the full-spectrum lights will last for nearly a year, burning night and day.

But you shouldn't subject plants to continuous light. Even when forcing spring plants so they will flower, the custom is to give them two nights in twenty-four hours by turning on the lights for an hour at midnight. In this way you get short nights, as contrasted to continuous long days (though the two are two ways of looking at the same phenomenon). Chrysanthemums, it is known, bloom only in long-night periods, usually October and November. To get them to bloom, darken your room for some extra hours in months that have longer days than those we get in October.

It is also interesting to discover that glass does not transmit sun rays of certain lengths, and that you need supplementary wavelengths to simulate what the plants would have received in direct sunlight. But leave it to the experts to do the experimenting with violet rays; it is dangerous to fool around with such things.

The radiant energy your plants must have is needed not only for photosynthesis in the green leaves, but also to trigger processes that control stem length, flowering, straightening out a warped seedling, and most enzyme actions. Though germination of seeds usually happens in the dark of the soil, radiant energy is needed for that process, too. All the energy used by plants (and animals) comes from the sun. There is no other.

A light source to use for your kitchen shelf—and under small growing plants in the garden, too—is a reflector. I use a strip of aluminum foil, sometimes crinkled, to put where the sun will strike it, and outdoors between the rows. Some people use even bigger reflector devices outdoors, and if you want this kind of aid for a partly shady place in your vegetable garden, make it in the winter. To see how well it works, try out reflectors on a sunlit shelf where you are growing lettuce, parsley, and cress. The heat will help the plants, too. Watch to see how good light affects your plants.

Plants ripen to full, blooming maturity in full sunlight (unless they come from the forest and are adapted to shade). In good light the nitrogen and mineral contents are properly balanced and plants conform most fully to their inherent characteristic type—with optimum root growth, shoot formation, and leaf development. Without enough light you can get them to advance, but they never have good quality because their proteins are arrested and many of their simple sugars do not develop into complex sugars. In this state they harbor the fungi you do not want, stretch and use up too much nitrogen to reach some light, or stay stunted with small structures, overactive enzymes breaking down the proteins and carbohydrates they do have, and using up their vitamin C. German studies of plant crystals that indicate whether the plants have been grown by organic or commercial chemical fertilizers also revealed that shade patterns and nonorganic fertilizer patterns are sometimes similar. The crystals of healthy, sunlit, organically nourished plants are alike and are a proof to Bio-Dynamic specialists that such plants are normal (and that chemically fertilized ones are not).

light . . .

Outdoors your garden will need plenty of light to protect it from dampness, slugs, snails, earwigs that love dark places, and the bacteria and fungi of decomposition that will go to work turning your plants into something for the compost heap long before you want them to. Even so, a lot of plants will grow, for photosynthesis goes on in almost any light—even moonlight. Some West Indians cut fence posts in the dark of the moon when the plants are lowest in sugar, and thus least susceptible to fungus attack and rot when put in the ground.

plants require certain temperatures

Combined warm and cool temperatures are preferred by most plants, as you'd naturally expect—warm in the day, cooler at night by about 5 to 10 degrees. Most of the house plants, however, that we have acclimated to our heated houses and apartments can stand a fair degree of heat at night. Anyone who has nursed along a cyclamen after Christmas knows, I'm sure, that a good cool place at night is the only thing that will save it—aside from correct watering, of course. The total range for plant activity is only 60 degrees: none to speak of below 40 degrees and none above 100 degrees.

how to stretch a plant's endurance

If you are growing plants in a northern climate in the winter, you are already stretching a plant's natural endurance. Outdoors, too, there are a few things to be done, and you can make preparations all year round. In the fall, when early *stretching endurance . . .*

frosts threaten, have protection ready to lay over the plants. In our town during a cold snap in October you can see old sheets and jackets spread over people's favorite tender flowers and vegetables when you go out early in the morning on the way to school or for the morning mail. In the spring, after a late frost you can see people out before sunrise hosing down their very tender young plants to melt the ice crystals that may have formed. I do this at least twice each spring to save the tender shoots of asparagus that are up.

Devices for providing artificial heat have been used for centuries, and in our age electrical gadgets with some new wrinkle come on the market every year or so. Many of them are very helpful in giving plants conditions that they prefer. For kitchen experimenting you can have trays and flats with small insulated electric cables set in down in the dirt. You can have any size of greenhouse from twenty inches up. You can build hot beds and cold frames outdoors, attached to your cellar window or up against a south wall. You can have large plastic bags to fill with warm water to warm up the soil when you lay them on the ground in the sun.

a heated flat for your kitchen

Any size of flat or planting box or shelf can be given heat by using heat tapes or cables supplied by electric companies and seedsmen. Never let them crisscross; lay them in loops, fastened in place by asbestos tape. The lining of the flat can be wood, metal, or polyethylene, and the first layer can be vermiculite, a very absorbent natural mica material available at hardware stores and garden centers. Then put in a four- or five-inch layer of soil, mixed with a moisture-holding material such as peat moss, kitty litter (or burned clay), or the moss from true bogs called sphagnum moss. Some seed companies sell the whole unit, all fixed up for you and ready for immediate use. (It is best to have a plastic or glass top of some sort to hold the moisture in the air near the plants. The unit from seed companies often comes equipped with such a top.) The advantage of warm soil for seed germination and young plants is that the soil solution which provides nutrients can be kept at the optimum temperature for good growth. If, however, it gets too hot, set up a small fan; and if it gets moist and sticky, a hair dryer will do the trick.

hot beds and cold frames

In the suburbs or in the country, you'll have plenty of room outdoors for a hot bed or a cold frame. The advantage of these garden aids is that you can stretch out the seasons by having plants started early in them, and can even grow certain crops in them for harvest directly from the frames. Plan to face either of these frames to the south, with protection on the north and windy side, but not under dripping eaves. Give them good drainage.

To build a wooden frame, make a structure that is no wider than 6 feet, no longer than 12 feet, and with a depth of 24 inches at the back and 15 inches at the front.

Do not try to make it wider than 6 feet or you can't reach into it to work in it. Do not attempt a length beyond 12 feet because then you'll have difficulty controlling the moisture and temperature, and consequently any pests which might come your way. There is no use enticing them to thrive in damp warmth if you can avoid it.

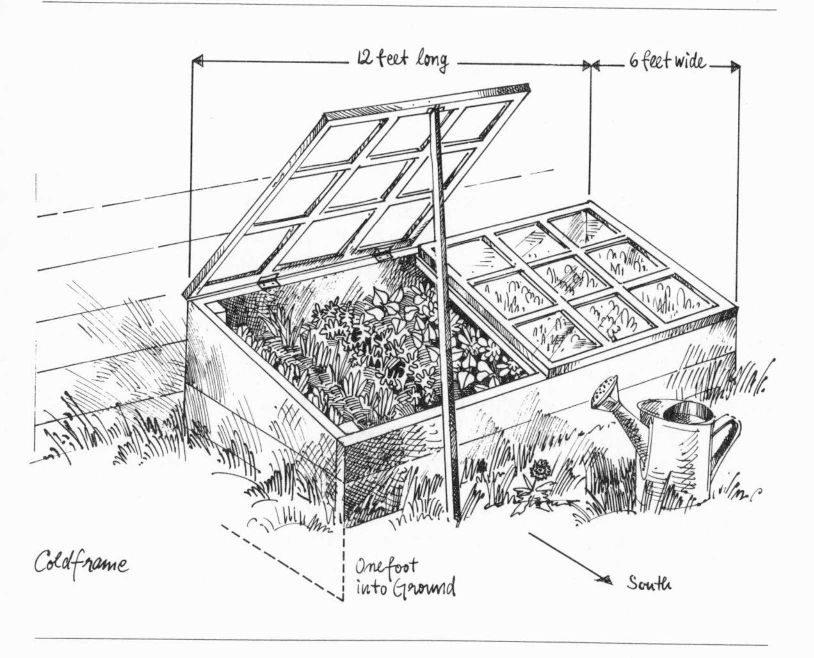

In the bottom of the frame, for a hot bed dig a 2- or 3-foot pit and put electric heating coils across it, or some hot water pipes if you build up against an apartment wall, house, or greenhouse with hot water heat. Put manure on top of the coils, then add a soil made of two parts loam, one part sand, and one part compost and/or peat and leaf mold. The sand will be of the best quality if you ripen it in the compost heap for a month or so. In perfectly natural *hot beds . . .*

hot beds only manure is used to heat up the mass, and the electricity (as with our grandfathers) is omitted.

Both hot beds and cold frames must have glass or plastic lids, of a size you can lift up to let in the air if the space inside gets too hot. The usual old-fashioned lid was a window sash, and two of them add up to about the right size for one bed. Put them on hinges so they can be easily lifted and propped up.

The advantage of a cold frame over a hot bed is that you can plan to raise plants out of season or let them linger out of season in an environment that is not different enough from the outdoors to let them get tender. In fact a cold frame, if used properly, will develop especially rugged and healthy plants. You leave them exposed to the weather whenever they can stand it, and protect them only when it is too cold for them to endure frosty nights or cold windy days.

The length of a cold frame is not so important as in the hot frame because you will air the cold frame more frequently and the moist warmth will be avoided for any extended periods. As for the location, it is even more important with a cold frame to use the warmest spot of protected southern exposure you can find. The only source of your heat is the sun. Over the frame you place window sashes or some sort of fiberglass. But never let it get hotter than 80 degrees. Close the frame at night to preserve some warmth and prevent frostbite.

Fill the frame with good rich loam, and plant thickly. Rows do not have to be farther apart than an inch or so, for you will watch their growth with an eagle eye, and pluck out all plants but the sturdiest anyway.

There will be rapid transpiration once the plants get started, so water daily, and use only tepid water. Little plants hate to be assaulted by cold water. Therefore, do not use a hose. Use a sprinkler with very fine holes at the end of the spout.

Another advantage of a cold frame—if you are interested—is that you can send away for plants that do not belong in your climate ordinarily and create a climate in which they might survive. Keep the temperature, in any case, at around 70 degrees.

If you plan to have both a hot bed and a cold frame, separate your vegetables into those that like a cool beginning to their lives as plants, and those that prefer it to be warm all the while. Some with warmer proclivities are tomatoes,

. . . *cold frames*

peppers, and eggplants, so they go in the hot bed. Those liking it cold include the brassica vegetables: broccoli, cauliflower, and cabbage. The unfussy ones like lettuce and radishes can be started early, too, in either the hot bed or the cold frame, where you can help them to get rugged and used to the cool of the outside air every day when you open up the frame. The cold frame can also be used as a transition place to help harden the young plants for final placement in the garden. You can start tomatoes or cabbages in the hot frame, then move them to the cold frame, and finally to the out-of-doors. For this much moving, you can plant your seeds in Ferto-pots or some of the new biodegradable pots or flats, so you will never have to disturb the roots while they are growing. This adds to your expense, but it is a fairly safe method, especially if you use sand in the mixture to help drainage, and sink the pots in loose earth to prevent their drying out. If you tend to overwater to compensate for drying out, use clean clay pots, to which little roots will not cling, for a safeguard.

If you do not use pots, but just move the plants from one ground to the other, let them establish themselves for several weeks. You will have broken off and disturbed the root hairs, and they need that much time to readjust. Be sure to water both the old and the new beds before transplanting, for this helps the little plants to survive the shock.

Organic gardeners who want to feed themselves and their families during as long a growing season as possible will use hot beds and cold frames as one way to stretch the seasons. The main danger, usually avoided by careful watch, is that of losing your plants to the fungus organisms called damping-off, which thrive only in moist warmth. Keep the air moving and you won't suffer from them too much, but if you see your plants drooping, or the stems getting rotten at the ground line, thin them out. Also put on a mulch of vermiculite, or even kitty litter.

When you start exposing cold frame plants to fresh air outside, do it slowly. Reduce the usual amount of water for a day or so before you open the lid for a full half hour. Then open it for a whole hour, then two hours, and so on. There will be wind, so your watering schedule will have to vary. In a week, you can probably leave the lid off all day.

In really cold weather cover the frames with blankets. When the sun comes out, take them off again, and let *hot beds . . .*

the sun get in, but keep the lids on. Plastic bags of warm water put on the soil in the morning will help it to warm up more quickly after a cold night.

Some people also use their cold frames for storage bins for such root vegetables as beets, carrots, turnips, and rutabagas. Even celery can be kept between good layers of straw. Remove a foot or so of soil before putting the lower layer of straw in for all these vegetables. Cover them well, inside and outside the frame. Invert a bushel basket filled with leaves over the latch so you can get at the vegetables under the snow, or when everything is frozen up. If you don't have a cold frame, half-bury an old refrigerator with the door side up, but be sure to remove the latch, of course.

One plant protection device you can plan for is an extended homemade Hotkap or cloche—which is really a structure like a long, narrow small greenhouse that you stretch right down the length of the row. You use curved plastic and curved wires like croquet wickets. Make the cloches in short sections, however, so that they can be removed easily if it suddenly gets too hot and muggy inside.

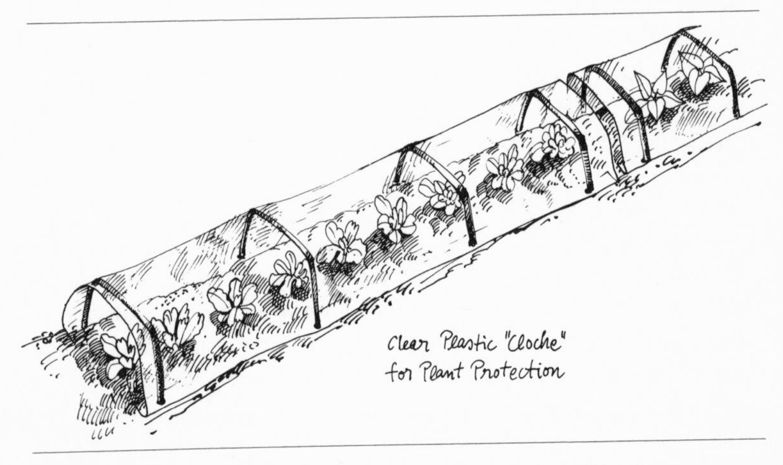

Clear Plastic "Cloche" for Plant Protection

Three-by-five-foot sections are feasible for a structure one foot high. Plan for wadding to stuff the ends on cold nights. Little plants do well in this miniature greenhouse when handled right, and you can expect corn or whatever you put under it to come along days ahead of the corn you plant out in the open.

A window bubble made of plastic can make a tiny hothouse out of your cellar window. An advantage of this is that instead of having to lean over, you can do your winter gardening right out the cellar window. When you want fresh young lettuce on a cold day, you do not have to go outdoors and walk around to a frame to get it. The other advantage is that when the hot frame garden needs airing, there is a safe source of cellar air, neither too cold nor too hot, to ventilate the plants. Some people buy window greenhouses to use out of upstairs windows.

. . . cold frames

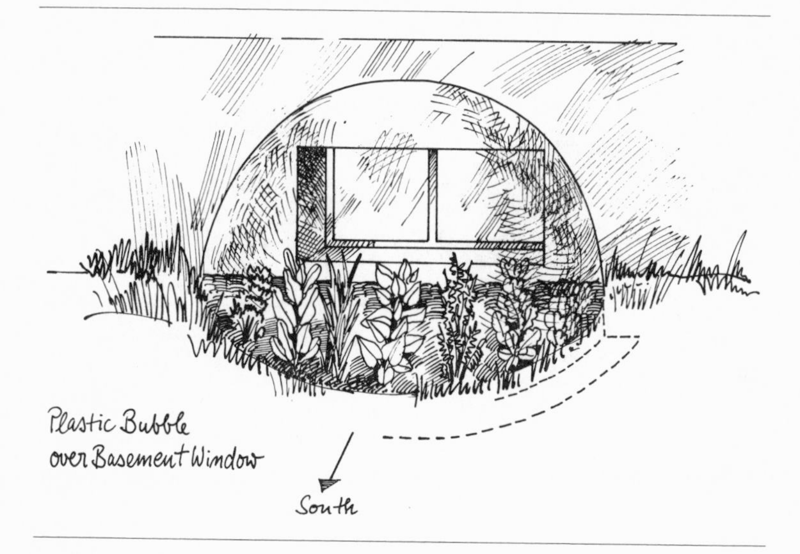

discovering plant structures

When you get your equipment all set, and your plants begin to grow, you find yourself more and more concerned with what is going on right under your own eyes, and curious about what plant structures are and what they are doing all during the cycle from seed through growth and flowering to seed again. As you become familiar with the way plants do their jobs, you get a sense of what they require and develop a skill in giving it to them if you can. Most important, you become aware of the entirety of a plant and of its earth, air, warmth, and water environment. You get to know all its patterns and rhythms, and this is surely the secret of the green thumb.

Keep a magnifying glass handy and look closely the minute your curiosity is aroused about something. Keep a very sharp knife nearby, too, so you can make a crosscut to examine. You might one day become curious about a potato—what it is, what it is made of, what function it serves on the plant, what its eyes are and what its sprouts are like. Take one out some evening and start slicing across it to find its structural parts. If there is a microscope in the house, all the better. Find the outer skin, the cork just under the skin, and then the storage cells inside with the starch grains, *plant structures . . .*

the parts of the eyes, and if it is getting toward spring, the sprouts themselves. Slice through a sprout and compare its textures and makeup with the big starchy potato tuber. Even ten minutes looking at a potato this way will give you an entirely different feeling from the one you have if you merely

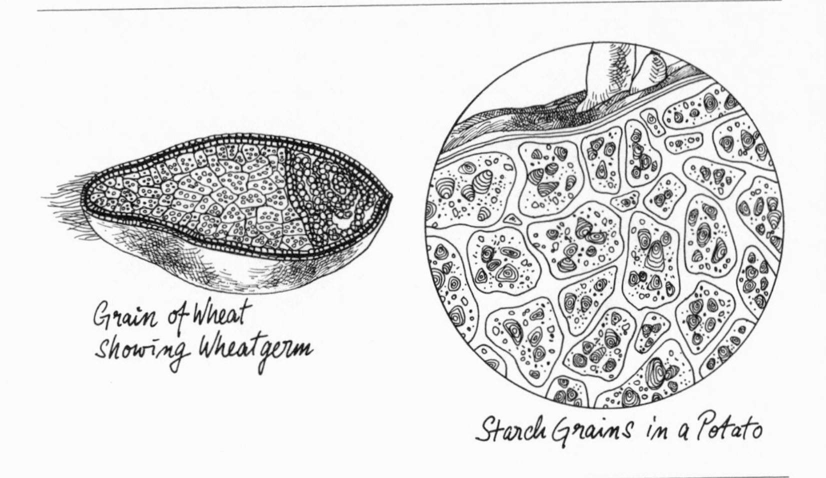

Grain of Wheat Showing Wheatgerm

Starch Grains in a Potato

look at it as something whitish with a brown peel to be cut off and thrown away. A potato tuber grows at one end of an underground stem, and is the storage place for the food which the big leaves of the potato plant will have been making all summer in your garden, or a good part of the winter on your window shelf if you decide to grow one as I did. The eyes and sprouts are for the next generation of plants that use the potato as storehouse for their food. Human potato eaters are invaders into that life cycle of the plants at the point of their storehouses.

how nutrients go up and food comes back down

You might also become curious about a carrot, or about how the nutrients go up and the food comes down into roots or tubers. Make a clean slice across a carrot and see all the layers. Way in the center is the woody part, in carrots as in trees. It is called *xylem,* meaning wood (as in *xylophone*) and is pronounced to rhyme with file′-em. This part functions to bring up water and nutrients from the soil solution in long cells which are structured so as to make it easy for liquids to pass up the root and stem to the branches, flowers, and leaves. These woody cells eventually become strong enough to hold up heavy plants. By the time trees are as big as redwoods, it is a tremendous weight these xylem cells have to sustain.

Next to the xylem is the *cambium* and then the *phloem* (pronounced to rhyme with Rome). The cambium layer is so thin you won't see it, but it is the vital layer, because from it are generated all the new cells of both xylem and phloem.

nutrients . . .

The phloem brings back to the plant for use and for storage the food that is created out of air and water in the green cells of the leaves by photosynthesis. Its cells are long, thin, and well adapted to their task. These minute structures are invisible to the eye, and quite hard to find under the microscope. But a good, well-illustrated botany book will clarify questions for you, and it will be well worth your time some long winter evening to study the way they look and work.

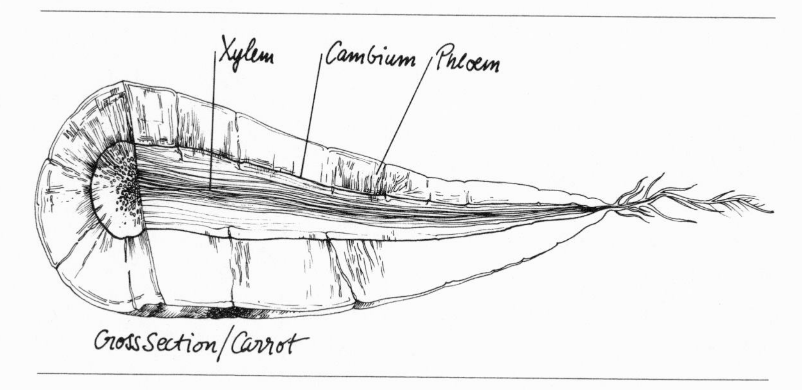

Cross Section/Carrot

It will be worth it, I think, because it gives you the intimate sense of nutrients going up and the food coming down in a perpetual rhythm. Afterward a plant can never again be something still, upright, and stiff to you, but a tremendously active, productive living organism. You are aware of how sturdy it can be, what enormous jobs it can do, but also of how fragile it can be and how injury to its infinitesimal parts might upset or ruin the workings of the whole living plant—and then perhaps a whole food chain. For you also begin to feel how the life of this plant is interwoven with other lives and with other structures in the environment and in the soil.

the one and only basis of all food

In the leaves of a plant all the green cells have chloroplasts, or little structures with the pigment chlorophyll, which makes the plant look green. Only when chlorophyll and light are present can the marvelous event of photosynthesis take place—when the plant transforms carbon dioxide from the *basis of food . . .*

air and water into simple sugar. That sugar becomes the food for the plant and thus the basis for all other food there is. It is the most important chemical reaction in the whole world, and is so simple and clear as to be miraculous. Isn't it astonishing that the energy of the sun itself can be transformed into the basis of the whole earth's energy in these tiny cells? Because it is happening in so many billions and trillions of green leaf cells all the time, we get blasé about it and forget that this single first step leads to all food and all energy available to us.

An oversimplified form of the equation is: $6CO_2 + 6H_2O + 674 \text{ kilogram calories} \rightarrow C_6H_{12}O_6 + 6O_2$. You can see that what happens is simply that carbon dioxide plus water plus calories react to yield sugar and oxygen. The oxygen produced with the sugar goes partly to the air and partly to the plant. It has come only from the water.

miles of surfaces inside leaves

The cross section of a leaf, if we could see it, is rather astonishing, too. On top is a protective cover, then packed cells, then spongy, openwork cells, and on the under surface holes like mouths to take in the air. In photosynthesis in the leaf, the water that combines with carbon dioxide comes up from the soil in a steady column, the carbon dioxide comes in through the openings, and the calories used in this

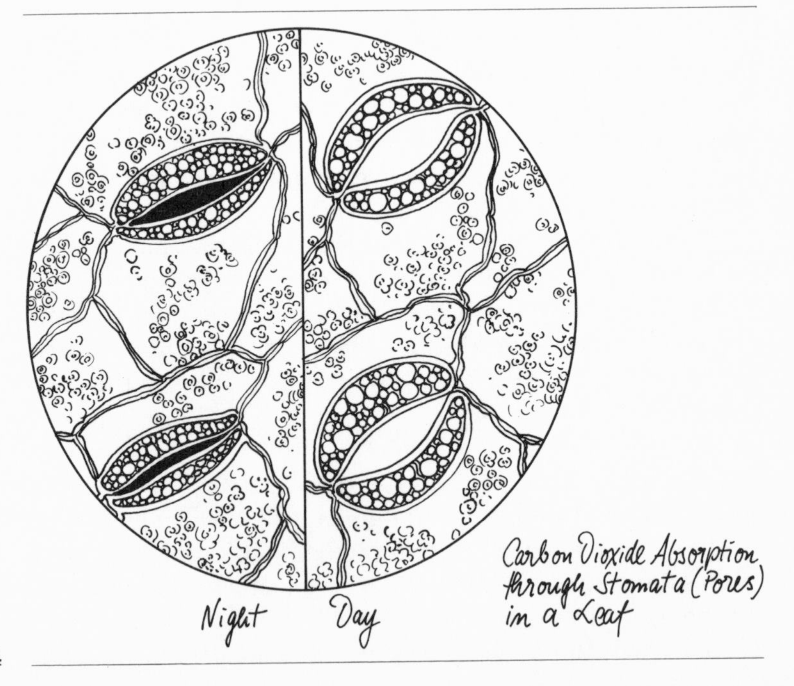

Carbon Dioxide Absorption through Stomata (Pores) in a Leaf

process come from the sun. In the layers of cells in the leaf there are millions of chloroplasts, and their surface area in a mature tree can add up to as much as 140 square miles. With all of those at work making simple sugar, that is quite a production. By the time the energy has been converted to animal and then human food (if you eat animal flesh), the energy has been considerably dissipated.

inside leaves . . .

The veins you see in a leaf are the vascular bundles, all having both xylem and phloem passageways. The cells that store food made by the plant are simple, thin-walled cells found in the tubers, pith, fruits, bulbs, flowers, and unspecialized parts of stems and roots. The plant itself is the best storage plant ever developed. I think it is up to gardeners to take advantage of this and learn how it all happens, and then to try to use plant foods at the height of their food value. Plant foods include, besides the sugars and starches, big protein molecules, nucleic acids, oils, enzymes, and such nutrients as iron, potassium, sodium, sulfates, and others. Plant cells also have pectin, tannins, cellulose, and lignin. You may have no urge to look into all these, but the very list suggests to any gardener the complexities of the biochemistry of plants.

tampering

Nutrients and growth substances of many kinds are available, and you can try some of them out on your plants to see what happens. If they are touted in ads as miracle substances and guaranteed to make your plants gigantic, and are based on newly invented chemicals or newly extracted hormones, enzymes, and fungus concentrates, the organic gardener is likely to be somewhat suspicious and refuse to fall for the latest appeal to his ego and laziness. If the new products tamper with hormones, he is likely to remember ugly warped plants and what has been injected into the necks of chickens in recent years to plump them for market, and he revolts. There is plenty of evidence that upsets of organic processes in the outdoors are dubious, to say the least. In an experimental laboratory, however, there is justification for using growth substances and other materials that affect plants. You can learn something from controlled comparisons of treatment, and you are not letting loose something into the environment if you confine your activities to a lab or a kitchen shelf. Of course you use caution and burn anything that might contaminate birds, animals, the soil, or the creatures of the soil.

tampering . . .

There are nutrients accepted by organic gardeners which you can use to see how your plants respond. These would include fish emulsions, seaweeds, cottonseed or other high-nitrogen meals, granite dusts, and rock phosphate. But again, for indoor purposes of comparison I see no reason why so-called complete commercial mixtures of nutrients containing set proportions of nitrogen, phosphorus, and potassium supplies should not be used. Knowledgeable botanists who count themselves as organic gardeners sometimes use commercial fertilizers outside also, along with their mulches and composts, to achieve even bigger, healthier, and thus even more pest-resistant plants.

Growing plants in plain water, with nutrients added periodically and frequent flushing, is a method used in many labs. Sometimes you can adapt this method by using peat moss or kitty litter to hold the plants in place. The peat moss shouldn't need it, but the kitty litter after a good soaking should be stored in the dark for several days to cut off development of unwanted algae. Perlite can be used, though it can get moldy; vermiculite, also, though it can get waterlogged and become unhealthy for plants. These are both natural substances.

Auxins and gibberellins are growth substances which many commercial growers use, and they also use sprout-inhibiting substances. Seedsmen use organic and other fungus inhibitors, too, sometimes, and often shock the gardener by the bright pink color of the seeds when they come out of the packet. Other seedsmen, as you know, use mercury poisons.

The organic gardener is often fussy about using any of these supplements—whether or not they are called organic. He wants to be a purist, perhaps. A supply of dried manure or compost from his heap or from the store, or the liquid made from steeping it for several days, is more to his liking. Instead of using any other fungicide he'll put a piece of charcoal into the water he uses (or bone black or activated carbon used in cleansing processes). Charcoal is very porous and absorbs the impurities, as well as oxidizing some of them. Fertrell and Sea-borne are organic nutrient supplements he knows about and trusts. Other nutrients such as foliar sprays might or might not affect beneficial bacteria and other soil organisms and their interdependencies, which we don't yet understand.

begin to do soil testing

To give yourself a good start in learning what you need to know about the biochemistry of the soil and plant growth, I recommend that you do some soil testing. There are several reliable firms that provide soil-testing kits, of varying prices depending on how elaborate you want to be in your search for the exact qualities of your indoor and outdoor soils. There is one that you can get for less than ten dollars, which not only gives you a chance to find out precisely how acid or alkaline your soil is, but also to see what you may need by way of nitrogen, phosphorus, and potassium—the three main additives in commercial fertilizer, and frequently the three most needed elements in the soil. Another company even provides a little guide to organic fertilizers, and a key to what your various vegetables might need by way of nutrients (see Appendix). You can practice on your indoor soil and if you can get hold of a six-inch-deep slice of your outdoor soil in the wintertime, start testing and start making plans for outdoor composting and for your supplies of materials to improve your soil.

These simple testing kits provide a dye which, when mixed with a sample of soil, will indicate on a color chart that goes with the set whether your soil is acid or alkaline and where it is on the pH scale. A soil might test from pH 3 at the very acid end to pH 7, which is neutral, to pH 11, which is at the very alkaline end (actually the scale is 1 to 14).

what pH means

What is being measured is the hydrogen ion concentration and its effect. When the molecules of water in the soil break down to their components, they have electric charges that attract them to other chemical components, enabling them to join to form new compounds. The positively charged hydrogen component, or ion, is balanced with the negatively charged other part of water, called the hydroxyl. At pH 7, the balance is perfect. When there are more positive hydrogen components, the soil gets acid, and the pH goes down from 7 to a lower number.

On the pH scale, the jump from 7 down to the more acid 6 is not just a one-unit jump, because it is figured so as to make it represent ten times as many hydrogen components as at 7. At pH 5, there are a hundred times as many. At pH 8 there are ten times as many alkaline units, and at pH 9, a hundred times as many.

pH . . . When there are a lot of hydrogen components on the loose, you counteract the acidity by adding lime to bring back the alkalinity. When it gets alkaline, you add sulfur and lots of organic matter.

Most vegetables like a neutral or slightly acid soil, with pH 6 to pH 7.5. This is exactly the range in which your soil organisms do best. Fungi take over below pH 5.5.

The common range in the mineral soils of humid regions is pH 5 to pH 7, or moderately acid. In more arid areas it goes to moderately alkaline conditions. Peat soils are very acid, with a pH going down to 3, and the extremely alkali mineral soils go up to pH 9. All this is important because the degree of acidity or alkalinity affects the soil solution and the capacity of the plant to use it and its nutrients.

If you experiment adding materials to your soil to make it more acid or more alkaline, you can discover eventually which specific plants like which specific conditions. A lot of study has already been done on plant preferences, however, so you can have the knowledge right away that a range of pH 5 to pH 6 is compatible for growing corn, soybeans, or tomatoes. A range of pH 6 to pH 7 is all right for asparagus, beets, and cauliflower—and even if it goes as high as pH 8 those plants will do quite well. Spinach, lettuce, peas, cabbage, and carrots also do well at this range from pH 6 to pH 8. These tolerances of the pH concentrations on both sides of the neutral point show that the practice of blindly adding lime to soil, without testing to make sure you need to raise the alkalinity, may well be a waste of time, though soil structure can be helped.

Gardeners who use a lot of organic material, however, should check often to see how their soil is affected in various parts of the garden. Also remember that bacteria do not like it too acid; their range of tolerance is only pH 5.4 to pH 9. Fungi have a range of tolerance from pH 3 to pH 9, so they can tolerate more soil conditions than bacteria can. Since you want both in your soil, you should take care to see that the range is right.

If you don't want to do any testing yourself, you can send a sample to your state university through your county extension agent. Test results can vary considerably, for conditions in your soil can vary within inches. The degree of moisture or dryness can bring on different results, also. Directions with soil testing kits usually recommend that only dry soil samples be used.

pH . . .

When you learn of an acid condition, try to find out its source. Leaching or washing away of soil nutrients may have caused acidity to build up. When alkaline bases dissolved into the soil solution by its acids are washed away, acids are left behind. The common bases are calcium, magnesium, potassium, and sodium, released from the minerals in the soil, exchangeable cations, and available for absorption—usually into the organic colloids and clay particles which are the intermediaries, as it were, between the soil and plants. Plant nutrients reach the plants via an interchange with the ions through the root hairs. A root releases one ion for each one it attains. It is the ongoing, perpetual interchange which makes the life of the plant possible.

If you learn that your soil, as tested, is rather alkaline, try to find the source of that, too. If, for example, the condition comes from chlorides and sulfates of sodium, calcium, or magnesium, you can flush them out. Your county agent can help you about these complicated matters when you happen to run into them.

To make a soil more alkaline, you add lime. The preference for the organic gardener is finely ground dolomitic limestone, as will be described later. To make more acid add pine needles, peat moss, sawdust, or tanbark.

what the nutrients are

Though you can't discover all the intricacies yourself without professional help, your gardening and windowsill experiments will hinge, nevertheless, on the fact that plants—as far as we now know—need fifteen of the elements out of the sum they contain.

These are: carbon (C), hydrogen (H), and oxygen (O) from the air and soil water. Also there are nitrogen (N), phosphorus (P), potassium (K), sulfur (S), calcium (Ca), magnesium (Mg), and iron (Fe)—all of which have been known since the end of the nineteenth century. In more recent years there have also been discovered as necessary: boron (B), copper (Cu), manganese (Mn), and zinc (Zn)—the trace elements. The last to be discovered was molybdenum (Mo). Plants also take in silicon, chlorine, and sodium, but it has not yet been shown that they need these elements.

Oddly enough, plants also cycle small amounts of a group of elements that seem to be truly nonessential: aluminum, arsenic, lead, barium, mercury, bromine, tin, cobalt, gold, nickel, and selenium. Some of the unknown causes of plant vigor might come from one or another of these, though the literature of organic gardening usually attributes that vigor to an as yet unidentified vitamin or two or three.

Animals need 14 of the plant's essential elements (it is believed) plus three others: sodium, cobalt, and iodine. Almost all of these go into the root of the plant in the ionic form, but nearly all can also go in through the pores in plants' leaves, either as a gas or in water.

Most nutrients are readily available for your plants in slightly acid soils, though some become unavailable when the soil is more than slightly acid, such as the one which is essential to the symbiotic nitrogen fixation carried on by soil bacteria. Phosphorus is an element in all soils, and unless the soil is nearly neutral it also becomes quickly unavailable. Potassium is very slowly available, but it is plentiful. Calcium, a base, is very easily washed away in humid climates, and the soil then becomes still more acid.

One good thing to do in the house to become versed in these facts is to supply a nutrient mixture deficient in one element, then to see whether the plant gets scrawny, spotty, yellow, or leafless. After that give it a foliar spray to observe the recovery when the deficiency is provided through the leaves.

Normal uptake of nutrients depends less on what the plant actually needs at the moment than it does on the rate of root growth and the chemical composition of the interchange surfaces of the soil particles. In fact, nutrients in amounts more than needed can go into the plant and sometimes even go in at a rate that turns out to be toxic.

If it is nitrogen you need, you might try urea if you apply it as a foliar spray, but it may burn because of the high osmotic concentration of the spray solution. The juice goes out of the cells when you want it to go in. The organic gardener is likely to try cottonseed meal or blood meal on the soil instead, even though he knows that urea is made from the natural nitrogen in the air—usually souped up with an additive, however.

If your plant needs sulfur and you live in the city or suburbs, just put the plant outdoors for a while if it's not too cold. It will pick up about what it needs probably from the sulfur dioxide pollution in the air. If it is magnesium you want, use an Epsom salt solution.

water

In addition to learning about nutrients, you will want to discover how water affects your plants. The best thing to do is get or make half-inch cotton wicks so you can study the movement of water in the soil and plant. Most people know very little about watering, and I've heard that more house plants are ruined by poor watering habits than by anything else.

Some people just poke the soil in a plant pot, or flat, to see whether it feels wet. This is a poor practice, for two reasons. First, daily pressure will eventually compact the soil and make it hard and airless. It's as bad as tramping around on a wet garden. Second, though the soil may feel dry, and actually be dry on top, it may have plenty of moisture below for the roots of the plant.

In our overheated houses, we can expect very rapid dehydration of the surface exposed to hot air, so enclose your plant in plastic sometimes for a day or so after watering to give it a chance to live a while in a nice moist atmosphere. Most plants would like that. This trick is used by people going away for a weekend so their plants won't dry out, but the wicks recommended above can be used as well. For weekend watering, just push the wick into the soil at the top of the pot.

To study water uptake, rig up the wick so that one

Test to study Water Uptake

water . . . end of it goes up into the soil at the bottom of the pot or flat, and the other end is in an enclosed jar. You can run the wick under the jar lid, and loosen the lid to a point where the osmosis of water is not cut off. Then tape the edge of the lid where there might be an intake of air extensive enough to cause undue evaporation. In general, all the evaporation should be absorbed by the wick itself.

Measure the amount of water you put in the jar, and then note the time. The point is to see what the rate of absorption turns out to be within twelve, twenty-four, and forty-eight hours, for instance. It is best to set up several plants, for comparison. And of course most interesting of all is to have one or two of them under plastic, so humid and dry environments can be compared too. You can vary your study by turning on a fan, an extra heater or hair dryer, an air conditioner, a humidifier, by running the shower or hot water faucet, or just by adding a lot of people breathing in the room. My botany teacher used to keep the plants in his greenhouse very happy just by turning the spray of cold water in the hose onto the hot steam pipes when he came in every morning. The humidity in that greenhouse was simply delicious.

begin composting

Ever since Sir Albert Howard invented the Indore method, composting has been the main clue to organic gardening. When you start on the fascinating processes of composting, you may become a totally converted, and purist, organic gardener overnight. Here are some ways to get started.

If you live in a cold climate, and really start in the middle of winter, the place for a bin or box is in the cellar. Because at certain periods there will be a fair amount of juice involved, it is better to use a wooden box and not paper. For safety's sake, you may want to line it with layers of newspapers or with a sheet of polyethylene. Then add a layer of sand, or a layer of peat or sphagnum moss. These serve both for absorption and for materials to help build the compost itself. Compost is old vegetable materials and other old organic materials which have been worked over by microscopic organisms (mostly bacterial plants) until they have turned into a rich, dry, black material. Humus is the same thing, but we usually think of humus as made in the woods and compost as helped along by the hand of man.

Into this box in your cellar you put all the vegetable residues left over from preparing your meals and from the plates afterward. Actually this is all you need to start with. There are additions which can help things along: leaves, dirt, manure, blood meal, ground lime, and much more. Some people like to add an activator to get the bacteria in there and working. Details and explanations can be found in the chapter about compost. For aesthetic reasons, you will probably want a lid on this box. In such a box or bin the work will be done by aerobic bacteria, those which need air, or rather oxygen, to do their jobs.

Anaerobic bacteria, or those not requiring oxygen, work in an airless place—best acquired nowadays inside a double thickness of plastic bag, such as those put in garbage pails. You can easily allocate one of your garbage bags to vegetable refuse, and carefully add such animal bits of refuse as hair, fingernails, wool, leather bits, and the like. Some people use meat scraps and cage cleanings, but I don't. There will be a sulfury, decaying smell to this accumulation at times, but it goes away. Organic gardeners put big bags of this sort on the back porch or down cellar, or hang them up in the garage. The fact that the bag is really airtight (and tie it tightly to see that it is) makes the possibility of an offending smell minimal—and only when one opens the bag. In fact, I kept some vegetable leaves tied tightly in a plastic bag on a radiator in my kitchen for two or three weeks last spring, and no one would have known it was there. No smell escaped until I took it outside to put on the earth. Then it smelled vile, but the juice and smell disappeared almost immediately into the soil.

This kind of back porch composting is a very simple matter. You just save your scraps, and anticipate the time that they will be converted to humus. A neighbor of mine in the cold climate of Vermont has a big family and a lot of vegetable residue and refuse to get rid of. She keeps a pile of big plastic bags of the 20- to 30-gallon size near the back door, and fills up one after another all winter long. They freeze, and are no bother to anyone. When the weather begins to warm up, she drags them out to one or another of her compost piles, and either lets the filled bags sit in the sun for a week or dumps them on a pile and adds blood meal, cottonseed meal, the ashes from her incinerator, sawdust, leaves, dirt, or whatever she has handy for the layers. She enjoys this back porch winter activity because, as she says, she has the "pack-rat syndrome." When she sees the telephone company truck go by in March with a load of wood chips after a tree-trimming operation, she runs out after it and begs the driver to bring his load in to leave near her compost piles. He always obliges, for it saves him a trip to the landfill site.

I know others who do indoor composting by having a barrel of earth in the cellar or back hall, and burying the vegetable refuse in that as it comes along. This contraption can be watered. You can also put in some red earthworms to help along the composting method by digesting their soil materials and returning their castings to the earth. This worm activity is one of the best examples of cycling you can watch, and everything used and returned makes the soil better and better. If there is any light where you have such a barrel, plant some seeds in it and see how lushly they grow.

Those of us who do winter composting feel we are quite a bit ahead with our gardening. Those who have the room and the inclination put their vegetable garbage on an outdoor pile. If out in the cold, bacterial action will be delayed until spring when the sun comes back. Then you will also add sawdust, leaves, and other material.

the crux of the plant

After all your experiments are done, and you have plants up and coming into bloom, take time to examine the crux of the plant, the flower. If it has been pollinated, fruits follow. Of course many of our vegetables are really fruits, unless they are seed pods, or seeds or leaves. (Most fruits are fruits unless they are pomes or berries or nuts.) Few people eat flowers, probably because of ancient superstitions or just plain ignorance. Candied violets have survived, but violets and nasturtiums in salad are rare on American tables. We eat cauliflower, of course, and the buds of broccoli. And an imaginative friend of mine serves chive flowers for their delicate onion permeation of a salad.

It is wonderful enough that we eat plants at all, and that their biochemistry contributes to our biochemistry. It is also wonderful that our ancestors tested them out, and found which ones were good for food, which ones good for medicine, cosmetics, emetics, purges, stimulants, poultices, and poisons.

If not often for food, we do take the nectar of flowers for perfumes, though nectar is mostly for bees and other pollinators. We don't take much, and it may have been a shrewd wisdom protecting the life cycle of plants that explains why ancient peoples held back from eating flowers. Plants or parts of plants that have been eaten as aphrodisiacs have had a very hazardous existence. Some in fact are extinct, and others are endangered species. The pre-Classic Greeks invented stories about the bad-luck effects of pulling up mandragora. By the logic of magic, its man-shaped root (called mandrake) was supposed to advance virility. The root would scream, the old wise men claimed. And that half-scared everyone to death just to think of. They left the plant to survive. I saw it growing on the Greek island of Delos, which was probably the crossroads of traffic in plant materials, even as it was of human materials in ancient days. Its green rosettes were evident in grassy hollows and among the anemones on the presently uninhabited island.

A flower is a means of facilitating fertilization and making seeds. It is bright-colored to attract the pollinators. The female part, in the pistil, has cells which receive the male cell from the pollen made in the male stamens. Then the eggs are fertilized, the plant fruits, and a new embryo can grow until we have new plants for beans, eggplant, peas, or tomatoes, all with new seeds for the next round.

It can happen in many ways. For some plants the agents are birds, bees, wasps, flies, butterflies, moths, or other insects (if you haven't already killed them with pesticides) who pick up pollen on their bodies as they move from flower to flower collecting nectar. Many flowers have the male and female parts both in the same flower. Other plants have separate male and female flowers, or even male on one tree or plant, female on the other. Some seeds blow in the wind. Some are carried on dog fur.

At the top of the pistil is a sticky or fuzzy or otherwise attractive and catching surface all set, at the right time, to pick up a grain of pollen for fertilization. Below it, reaching to the ovary, is a passage down which the male cell stretches its pollen tube to do the actual fertilizing of the ovule in the ovary.

In gardens drenched with poison pesticides a doleful pattern of events can be expected. There are no insects; no transferring of pollen; no natural fertilization; few seeds or fruits or flowers to develop to put in your salads. A year ago I saw a film of apple orchards in Japan, showing avenues of trees so saturated with sprays that all the bees in the area are dead and gone. Pollination of every apple blossom that was to set fruit in these orchards had to be done by hand by conscripted army personnel.

Often there are many eggs in the ovary, and a pollen tube that arrives releases two sperm cells. One unites *crux of plant . . .*

with an egg and commences the embryo—for example, wheat germ; the second unites with a structure that begins the growth of the endosperm, the sturdy food storage area.

Germination has been known about for a long time, though sometimes forgotten. Theophrastus, a pupil of Aristotle, pointed out that different seeds germinate differently and that the root is the first structure to start growing.

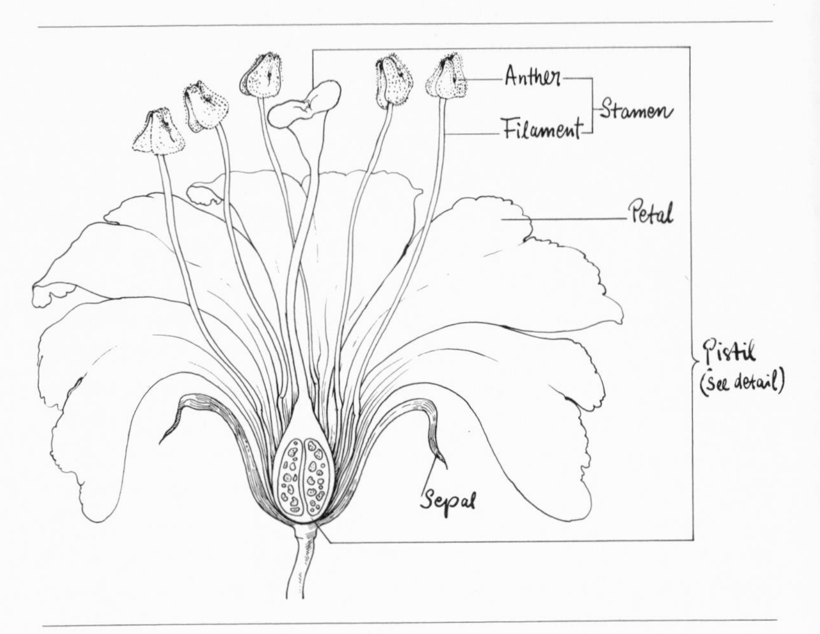

plant heredity

It was not until Gregor Mendel in the nineteenth century, however, that scientists discovered why plants should not pollinate themselves. To be strong and adaptive, an organism needs two different parents, with different genes. Insects, such as the Mexican sleeping sickness mosquitoes, that adapt to new hydrocarbon pesticides have shown how tremendous adaptive vigor can be. Horticulturists who hybridize for blight resistance have also demonstrated that new genetic strains can be stronger and healthier—at least for a while—than some of the previous popular garden varieties which may have been weakened through breeding or overspecialization somewhere along the line. This need for genetic vigor is one of the reasons that botanists and ecologists are so anxious to preserve primitive long-lived species which might turn out to be exactly what is needed to breed a new hybrid in times of bad crops or a corn blight plague. A gene pool of this sort was in great demand during the 1970–71 season when corn blight threatened the crops in the Midwest.

Farmers in Iowa that winter bargained and bartered furiously to get nonhybridized seed they felt they could trust.

exchanging information

The suggestions in this chapter—and dozens of other activities and preparations that will occur to you—are among the ways a gardener can begin to use time indoors to get used to the ways of gardening, get prepared to do it, and develop a feeling for growing things. I am convinced that the more ways of relating to plants and gardening that you can develop, the better.

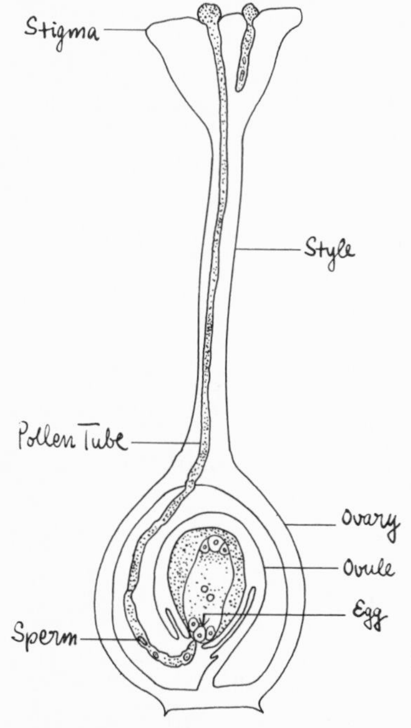

In our town (and it has happened in many others, I imagine) we started an organic gardening club during the winter months to exchange ideas, addresses for sources of materials, and the names of farmers who would sell aged manure, or let persons come to clean out the henhouse for a supply of fresh manure. In the spring we were all too busy to get around to visit each other, but in summer we had fine, fascinating trips to each other's gardens and got ideas about growing comfrey, for instance, if we never had—or about how to get a really superior crop of edible weeds, lamb's-quarters, among the beans (or beans among the lamb's-quarters) by a cavalier neglect.

We also swapped information about farm ponds, raising fish and frogs, and heard great accounts of how many earwigs a frog would eat if you'd catch them for him.

I don't mean to imply that all the suggestions I've made are essential to the would-be organic gardener, but I know they can add a new dimension to the life one can lead with plants. All I want to stress for the beginner is that there are three things of special importance to do: begin composting; begin to know the growing habits of plants by intimate acquaintance in your own house, in a spot where you can watch what is happening; begin to make plans, buy equipment, order seeds, and, as far as possible, practice ahead of time what you are going to be doing outdoors later. The next chapter goes into detail about planning.

chapter three

get ready, get set, plan

If you are already on the mailing lists, your seed catalogs begin arriving in January. It is a welcome change from other studies and activities to settle down on a winter evening to select which seeds you will need for your garden in the spring. Though it is an enormous temptation to get too much, and tricky to figure the right proportion of vegetables, flowers, and herbs to plant for good health and pest deterrence, careful planning will keep things under control.

But be bold. Plan to do some outside experimenting to follow your indoors experimenting. Select new herbs and vegetables you have never had before—or never heard of before. Try out some you have never thought you'd like, and see whether this food, eaten fresh, young, and tender, is actually something you never really tasted before. And be sure to include enough pungent, pest-free, or at least pest-repellent flowers and herbs to intermingle with your vegetables and to surround the garden.

When you have studied up on these matters (consult Chapter 6 often) make a layout. One is diagrammed on pp. 56–7, but you will have to figure out your own, of course, to suit the terrain and exposures you have, as well as the space.

plan your whole garden

During the long winter evenings you should map out your entire garden, not just the vegetable part of it. Always keep in mind that you aim to create a garden that is a total living environment—not only for vegetables but also for trees, shrubs, grasses, and flowers and for all the beneficial (and instinctively marauding) bacteria, fungi, actinomycetes, mice, shrews, moles, chipmunks, squirrels, skunks, gophers, and helpful frogs, toads, ladybugs, lacewings, and lots of birds who will frequent your place if it is well balanced and adjusted to the foods these creatures require. They keep themselves under control if given half a chance, so you never have an unmanageable population explosion of any species.

vitamins involved

If you approach your garden planning from the point of view of nutrition, four vitamins will be of importance. For vitamin A, consider such good sources as carrots, chicory, dandelion, string beans, peas, kale, parsley, spinach, turnip greens, broccoli, chard, collards, endive or witloof chicory, lettuce, peppers, pumpkins, squash (Hubbard), and sweet corn. For vitamin C plan to select any of the following: beets, broccoli, Brussels sprouts, cabbage, cauliflower, kale, kohlrabi, parsnips, peppers, potatoes, tomatoes, mustard, parsley, radishes, spinach, and sweet corn. If you live in the right climate, include also avocados, bananas, lemons, oranges, grapefruit, muskmelons, and berries like strawberries and raspberries. For vitamin B_1: green lima beans, Brussels sprouts, peanuts, green peas, potatoes, tomatoes, globe artichokes, asparagus, and beets. And for vitamin B_2: avocados, carrots, chicory, green peas, spinach, green lima beans, kale, and turnip greens.

Of all the vegetables, turnip greens, kale, tomatoes, and parsley are very high in nutrients. For healthful herbs try: anise, balm, basil, chervil, dill, marigold, marjoram, parsley, nasturtium, summer savory, sage, and thyme—all suitable for indoor growing also.

winter storage

To plan for vegetables you can store in pits or cold cellars, consider especially those you can sow in the summer and have at top quality to harvest in the fall. Plant late turnips, carrots, beets, cauliflower, cabbage, rutabagas, and also late cabbage, broccoli, celery, kohlrabi, onion sets, winter radishes, chicory, Chinese cabbage, winter squash and pumpkins, potatoes, and beans and peas to dry. If you have no place to insulate off in your cellar to make a cold-storage place for your winter vegetables, you should make plans for pits in which to bury them in moist sand under leaves. Others that can be stored include Brussels sprouts, parsnips, and such fruits as apples and pears.

The vegetables best to freeze include: asparagus, lima beans, string beans, broccoli, cauliflower, chard, corn, kale, peas, cooked pumpkin, spinach, New Zealand spinach, and such berries and fruits as blackberries, blueberries, cherries, currants, gooseberries, raspberries, rhubarb, and strawberries.

winter storage . . .

Those to can in Mason jars are peppers and tomatoes—don't forget tomato juice and grape juice—asparagus, lima beans, beets, chard, sweet corn, spinach, and such fruits as cherries, peaches, pears, and rhubarb. Under no circumstances can cabbage (except sauerkraut), cauliflower, celery, cucumbers, baked beans, eggplant, lettuce, onions, parsnips, turnips, or vegetable mixtures because of the danger of bacterial contamination. All precautions should be taken to avoid the deadly germ of botulism. If the directions in your cookbook say 75 minutes for sweet corn at 10 pounds pressure, leave them in the cooker for 75 minutes, not a minute less.

Vegetables to dry include lima beans, shell beans—right on the vine—carrots, corn, peppers, pumpkin, and winter squash. I used to have a little cloth and wooden contraption for drying vegetables, and it worked quite well, but I liked it better for herbs than for vegetables that could be easily stored in other ways. Dried fruits, because they are so delicious, are something else again. Try cherries, grapes, blueberries, currants, and pears. Plums, too, of course, and make your own prunes.

how to read a seed catalog

Send away for as many seed catalogs as you think you'll have time to read, but in general, deal with seedsmen in your own zone. The names of many seedsmen are given in the Appendix, but read the newspapers and garden magazines for other ads and addresses. Occasionally, a specialty nursery like Wayside Gardens, in Mentor, Ohio, or the White Flower Farm in Litchfield, Connecticut, will ask you to pay a couple of dollars, but they will deduct that amount from your first order. Send to different parts of the country—Parks in South Carolina, Farmer Seed and Nursery in Minnesota, Veseys in Canada, and to such old, established firms as Stokes, Harris, and Burpee. There are a few out-and-out organic gardener houses like Nichols in Albany, Oregon; Natural Development Co. in Bainbridge, Pennsylvania, and Walnut Acres in Penns Creek, Pennsylvania. In the organic-gardening books and articles there is not much stress on organically grown seeds, perhaps because of a feeling that the hard pesticides have not penetrated the hard seed coats into the endosperm of the seeds, so the seeds remain fairly clean. In the last few years, the established old houses have informed their customers whether or not their seeds have been treated, and with exactly what. Sometimes it is a mild fungicide like Captan; sometimes it is merely a heat treatment. If you ask a house like Harris for untreated seeds, they'll send them to you. I tend to feel good about a catalog that tells you just what the company does, and to trust that company to be reliable. Also, look to see what fertilizers are being advertised. If, like the recent catalogs of Wayside Gardens, the only ones they advertise now are organic composts and dried manures, again you feel good.

seed catalogs . . .

Carefully scan the garden aids to see whether the company offers sensible ones in accord with the gardening principles you are learning. In recent years, for instance, small inside table greenhouses have been advertised in the garden-aid sections of various reliable companies. One from Burpee's has a plastic cover which, if too tight, would make the little 14-inch house get moist and cause the disease of damping-off. But the ad says it has a ventilating adjustment, self-watering saucers, and an air space between the plant tray and the outside container, all excellent precautions. What's more, in 1972 the cost was low. A heating cable, of course, costs more.

Less reliable companies claim magical results with unspecified substances or unfailing success for exotic plants and unusual, freaky specialties that are supposed to appeal to your eye for a bright novelty, and not to your liking for solid gardening and the ordinary hard work it involves. The only timesavers and worksavers you want to fall for are mulches, some of the quick compost aids, certain harmless tools, and very rugged, disease-resistant, fast-growing plants and clean seeds.

Be wary of those advertisers whose catalogs have gaudy pictures of everblooming rose trees, hibiscus blossoms twice as big as a child's head, and blueberries as large as apricots. Why be enticed by promises about fairy-tale cascades of three or four kinds of roses, or nuts, or fruits on one tree, or potatoes and tomatoes on one vine?

Be sure to study planting zones and learn what can and cannot be grown in your area. It can be hard to resist temptations to grow some plant in your zone which is not hardy there, or suited to your climate or soil conditions, even though a company promises success. I am greatly tempted to try growing artichokes, though I know better. Yet, I persist in hoping some smart plant breeder will come up with a new hybrid which can endure the rugged conditions of Vermont. The rosy come-on of catalog copy usually includes appeals like "lavish crops" and "flowering glory."

Then it states: "Successfully grown as far north as New York," and my better judgment checks me because I know artichokes like a moist temperate climate, preferably not far from the sea, and I know specialists recommend artichokes as something to try only in seaside areas—not inland as far as we are.

The vocabulary of all catalogs is apt to be excessive in one way or another, but as a clue, look for quiet claims like "heavy crop" or "bears well for several weeks, if kept picked" instead of the glamorous promises. Look also for a full, logical collection of offerings, not a haphazard spotty one.

One sign of a good catalog is that the company gives you planting hints. Down in the corner of a page on corn will be a box telling you, for instance, how to grow sweet corn, how far apart, how deep, what kind of soil, how much a packet will plant, and when to pick. They do this for each vegetable. A company like this never advertises fertilizers as "instant action" or says "one spray kills all." And they never advertise Venus'-fly-trap or the magic carpet of flowers which you just unroll and keep watered. Or ceramic cats and toadstools. Or plastic grass.

Cheap trees and shrubs will be just cheap, though I do know organic gardeners who have such good soil and such green thumbs that they are able to remake them into valuable plants. A Russian olive at fifty cents, will, yes, eventually grow to be a fine bush, but at this price it will be such a spindly little thing when it arrives that you can hardly see it for several years in the hedge where you put it.

But there is one thing to say for cheap plants. They are often cheap because they are of a variety that does very well in all kinds of conditions, spreads fast, and is so rugged that the plants are fast multipliers, hard to control. Sometimes you can find bargains in pachysandra for ground cover, or other ground covers like lily of the valley and ajuga; or sometimes bargains in multiflora roses for hedges; or for sedums or vetch. One perennial plant called crown vetch is a winter-hardy legume that will fill up banks and gullies. Twenty-five plants will cover 100 square feet in three years and it may take you over in four years. For fifty cents a plant, in five years you would surely have your money's worth—if you want to be swamped, that is.

Sometimes these glamour-appeal catalogs do, however, have good gadgets that appeal to organic gardeners. The Klippety-Klop windmill or mole chaser, for example, is a good device, designed to oust moles by vibration and not by poisons. Another device, a cheap 12-foot frame with plastic over it curving to the ground, is not only a good adjustable cloche for a tender row of small plants; it can be used as a rabbit repeller, too. Other fairly good bargains are nets, strips of plastic for mulching or compost protection, small birdhouses and bird feeders. So are some of the hoes, augers for deep feeding, and staples like flower pots and hoses. Most of the reliable and trustworthy old houses sell these extras, also, of course.

When you settle down to look over the catalog of a reliable company, first scan their features of the year in the front of the catalog, but don't make any decisions yet. Just keep your eye open for suggestions particularly useful for you and your needs—such as a new bush variety of squash that will save you a lot of room, or a well-proven disease-resistant variety of pepper or potato, or a long-season one of lettuce which will increase your harvest, or a variety that gives you a better vitamin supply than any of the other tomatoes, for instance. Just take note of these and then compare the features with the extended descriptions in the body of the catalog, especially for prices. Sometimes there are savings advertised in the front pages. A special called "A Family Garden Collection," for example, may offer all the vegetables you would plan to grow, and in varieties with just the qualities you seek. It may be a very good bargain—and after the soil at your place has built up, you can count on your rich, composted, and well-mulched garden to make these vegetables yield well.

In the main part of the catalog you should read every direction and plant description, and also every box with special pertinent information. You will discover how much one of the packets will sow, general directions for planting, and special difficulties to be avoided. All too many catalogs will still recommend hard chemical sprays made out of chlorinated hydrocarbons. Learn to disregard all such things, and only to pay attention to the biological or botanical suggestions including rotenone, pyrethrum, and sabadilla, and other organic preparations. Then read every description under a certain vegetable to see what qualities the seedsman finds pertinent to stress, and judge whether he is addressing the commercial grower or the home gardener. Some may stress size and shape; others color; or firmness; or sturdiness and flavor. You can also see by the amounts offered (100 pounds, bushel, etc.) whether the seeds are intended for *seed catalogs . . .*

commercial growing. Think twice about such aspects as sturdiness or an ability to stay tender until large size. These comments hint at durable market quantity rather than home-gardener quality. If it is a variety of string bean you are choosing, for example, turn down a bean praised for an ability to stay tender. You already know that if you can possibly help it, you are not going to let your beans ever grow to full size and lose their flavor and natural tenderness. In a home garden you can pick all your string beans when they are one-half to two-thirds grown, and get all the benefit of young sweet beans. The same goes for peas. And these vegetables are much better frozen or canned at this age, too. (See Chapter 6 for further suggestions.)

the layout of my garden

At our semisuburban place we keep the main growing areas to the east and south of the house, not only because that is the area away from the road, but also because that part of the property has the right southern and southeastern exposure, as well as the right slope toward the southeast.

There are maples in the front yard and across the driveway from the house and also in back of the garage, but the fruit trees are toward the south and rear, arranged as much for the birds and their comings and goings in the backyard as for us. Trees especially chosen for the birds were Russian olive, elderberry, honeysuckle, highbush cranberry, and mulberry. We share the grapes and edible berries, and watch with pleasure the way the house sparrows gobble up the Japanese beetles on the grape leaves in August. The birdbath and birdhouses are placed in strategic spots, and we keep bird feeder tables under the apple and pear trees. The potted flowers and vegetables on the terrace and patio area keep the hummingbirds and bees coming.

Flowers and herbs are conveniently located, with good sunny exposures. Various herbs are sometimes transplanted to the vegetable garden as protection plants, either at the beginning of the season or at times of invasion and trouble. Wormwood, of the *Artemisia* family, is a quick worker to move to a trouble spot because of its powerful root exudate, which goes forth into the soil to fend off marauders.

The rather heavy planting on the northwest edge of the vegetable garden serves both as windbreak during spells of bad weather and as shade and protection for some plants which like that shade. The mulberry tree, though its roots do rob some plants growing near it, is a great attraction for birds, many of whom eat a few insects and grubs while they are in the neighborhood. This tree and the honeysuckles nearby serve as cover also, for birds like to have some place to hide if they are startled. The garage door is left open in the summertime for the barn swallows who come to nest there.

The compost heap is placed conveniently near the vegetable garden, and also not too far from the driveway, so that heavy things brought in can be wheeled right down. The toolshed part of the garage opens toward the garden and the compost heap, so equipment is handy. Beyond, to the east, are a field (with hay handy twice a summer) and a pine woods (good source of needles for compost and mulch).

The vegetable garden itself is surrounded by pungent annual flowers and perennial herbs such as mint, tarragon, and sage. The rows run across the slope, which fortunately also makes them run more or less east and west—though for good sun benefit during the summer north-south rows are also perfectly satisfactory. We live so far north that east-west rows suit us; there is a lot of morning east sun and afternoon west sun because the sun up here rises so far north in the summer.

We plant six rows of corn, and move that block around the garden in a three-year rotation; we also move around our four rows of peas, two rows of beans, one row each of carrots, beets, parsnips, and various lettuces. The tomatoes and squashes, planted on the edges near the herbs, are rotated around to different edges; this is not much of a rotation, so we plant protective plants quite thickly in their areas.

Of course your garden will be different. Consult your seed catalogs, read the packets, ask your county agent and all your friends for suggestions about the best amounts and varieties to use for your kind of soil and climate.

All estimates, of course, are subject to individual preferences, and the way you do your thinning of plants after they come up—and of course your thinning depends on the way you do your planting. If, eventually, you want your carrots two or three inches apart, so they won't twist around each other and be crowded and stunted, you thin them to this kind of spacing, and eat the little ones you pull out. (See Chapter 6.) You will have some temptation to plant the seeds in the first place at such distances, but not only is it

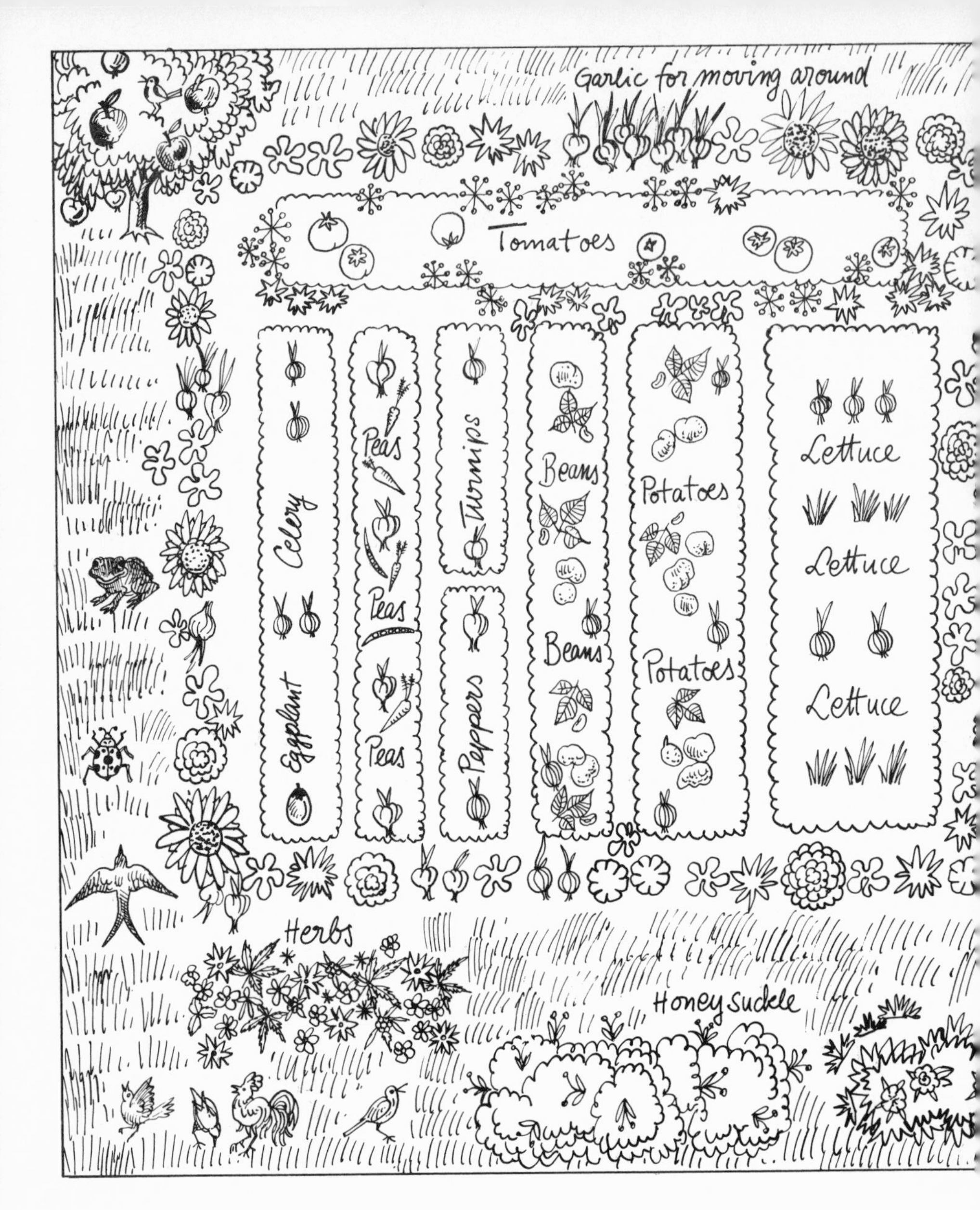

White geraniums (repel Japanese beetles from corn, asparagus, kohlrabi, soybeans)

Mexican marigolds (repel nematodes and control club root in brassica vegetables; protect cucumbers)

Radishes (mark rows and precede long-term crops)

Nasturtiums (rebuff cucumber beetle, Mexican bean beetle)

Garlic (rebuffs rabbits, many insect pests; protects apples, pears, cucumbers, peas, celery, lettuce, etc.; makes useful spray)

Onions and Chives (same as above but less potent)

Dill (trap plant for tomato worm)

Zinnia (trap plant for Japanese beetle; rebuffs cucumber beetle, tomato worm)

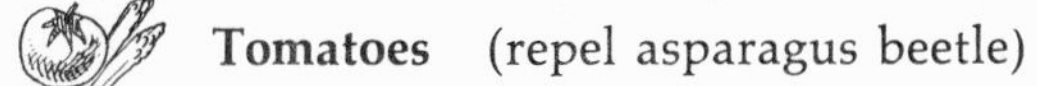

Tomatoes (repel asparagus beetle)

Strong herbs: sage, thyme, catnip, feverfew, hyssop, rue, artemisia, etc. (repel many pests)

Parsley (protects hybrid tea roses)

Ladybugs (devourer of many pests, including many larvae)

Peas and carrots thrive together, as do **potatoes and beans,** parsnips and peas, beets and broccoli, radishes and cucumbers.

Turnips (repel aphids, spider mites, flies)

very difficult to do with tiny carrot seeds, but young plants for some wonderful community reason like to be together. It may be that they help each other by the substances they exude from their roots.

herbs

Herbs are an essential element in the organic gardener's design for growing vegetables, for they serve several purposes and in addition are excellent to add to many dishes for tasty and subtle results.

In making your selections about what to plant, you want to pick out those that will benefit your garden in three or four ways. For the "herbs of general good influence," according to the Bio-Dynamic gardeners you would choose borage, to plant at one corner of the bed, lavender in a few clumps at another corner, blue hyssop at another, and sage at the fourth corner. In addition, for "good influence," you would include the equivalent of a row of parsley divided up and spread around into four- or five-part rows, plus chervil and marjoram. Also to be divided and spread around are tarragon (the small kind; see p. 171), dill (except near carrots), and chives. Two other plants in this category are camomile and lovage. The gardeners of this Bio-Dynamic school also stress the value of adding the residues of these "good influence" herbs in the compost heap. Whatever the influence is, it is doubtless a good idea to keep it cycling as you keep cycling the virtues of all the green plants you put on your compost heap, with their mineral content, vitamins, and other nutritional values.

The basic herb supplements for the compost heap, according to this school are: chicory (the blue-flowered kind), nettle (the stinging kind), camomile, dandelion, yarrow, and valerian. To these the Bio-Dynamic composter adds ground bark of oak for its calcium content. The virtues of the herbs are considered to be: chicory, an alkaline salt and silicic acid; nettle, sulfuric compounds; camomile, a power to regulate a plant's use of calcium; dandelion, a power to regulate silicic acid; yarrow, a power to guide the affinity between potassium and sulfur; valerian, a power to help phosphorus compounds to be assimilated.

Horsetail plants are brought to the garden and used by these herbalists for brewing a tea to make a spray to control the nonbeneficial fungi. Other gardeners dust with rotenone, but the principle of using the heavy silicon content

of horsetail plants is well worth remembering. (Slugs hate silicon, too.)

herbs . . .

For more information about Bio-Dynamic methods, see the Appendix for addresses. The magazine these gardeners publish is full of rousing little articles, exhorting readers to note the "deplorable conditions" which obtain, and the need for punitive measures against old-fashioned teachers, farmers, and agents who advocate practices of poisoning, and soil treatment which they consider ruinous.

The herbs, flowers, and weeds that I find most valuable are the pest-repellent ones. Some of them turn back insects from their approach to your vegetable plants; others distract them and, as it were, trap them there to consume these weeds or flowers instead of your vegetable plants. The insects are usually quite easy to see on the trap plants and pick off for dousing in kerosene.

For herb teas to use as sprays to protect your other plants from pests you will want plenty of garlic and some achillea or yarrow, camomile, St.-John's-wort, chives, mints, oregano, sage, and horseradish. Two or three of these added to hot pepper make effective sprays.

For attracting bees you can include thyme, catnip, lemon balm, pot marjoram, hyssop, sweet basil, and the mints. Lemon balm is so attractive that old beekeepers used to rub the inside of a new swarm with it to entice the bees to stay in the house. Also plan for some of the following for your hedges: bee balm, *Rosa rugosa,* elderberry, and privet. These will also please the birds, and for food value you will prize the rose and the elderberry.

Tansy, spearmint, and pennyroyal will ward off ants. Aphids (who are harbored and farmed by ants) stay away from spearmint, stinging nettle, and nasturtiums. Sage, rosemary, hyssop, thyme, and the varieties of artemisia known as wormwood and southernwood will help to keep the cabbageworm butterfly from coming near your broccoli, cauliflower, and cabbages. They will also turn back the black flea beetle.

Where there are infestations of Japanese beetle, people are glad to learn that white geranium and the poisonous *Datura* called jimson weed or thorn apple are repulsive to that beetle, and that zinnias and knotweed are good trap plants to distract them.

companionate planting

This is the term used for the intermixture of plants so that they in some way benefit each other. Mixing herbs, flowers, and vegetables as described above is one way. "Companionate cropping" is the term used for putting certain plants between others in the same row; "intercropping" is used to describe the practice of planting one kind of row between two others of a second kind of plant. It is beneficial to practice companion cropping by planting beets between cabbages; spinach between cauliflowers, or between eggplants or celery plants; and onions or any members of the *Allium* family in the row between beans, lettuce, or young cucumbers, which rabbits or woodchucks like to attack when young and tender and full of protein.

These combinations may also be used for intercropping, and others to plan for might include beans and potatoes to control Mexican bean beetles, eggplant, flax, or green beans to control Colorado potato beetle, chives among the roses to chase aphids, and marigolds intercropped with beans. Intercropping to save space can be achieved by planting spinach between rows of beans or peas on wires, lettuce between rows of cabbage or pole beans, early peas between pole lima beans, and spinach or radishes between squash, cucumbers, or pumpkins. The radishes have the extra benefit of fending off the striped cucumber beetles and squash bugs, too. I also use nasturtiums for these purposes, and my neighbor who redeemed her garden from a gravel ledge uses onions. If you select an orange nasturtium, and add some orange marigolds, you will perhaps gain an added protection, for aphids and other bugs are believed not to like that color and fly right past it.

Vegetables that benefit each other by being planted near together are peas with carrots; bush beans with celery; beans or corn next to cucumber; beets near kohlrabi; corn near potatoes; lettuce on both sides of radishes; celery with leeks. Parsley near tomatoes seems to help them; peas help turnips.

other space savers

Another space saver aside from companionate intercropping is the vertical or double-deck mode of gardening. This can be used on terraces and in small plots and roof gardens. What you do is grow pole beans, wired peas and cucumbers, and staked tomatoes, all of which will climb up and grow above ground. Then underneath you plant protective herbs, lettuce, radishes, spinach, and squash or carrots. To save more space you can choose any of the dwarf varieties: cherry tomatoes, Tom Thumb corn, and little snow peas. Also rotate your early, midseason, and late crops and use all the odd corners you can think of. Do succession planting and use tubs, boxes, and pots to expand your planting area.

space savers . . .

In making all these plans be sure to figure on high-growing vegetables to the north of your low-growing ones which need a great deal of sun. In fact, most vegetables, except lettuce at certain stages, do need a great deal of sun. Corn or pole beans are likely to be your tallest plants, so put them toward the north. Other tall vegetables include pole peas or wired peas, staked, highly fertilized tomatoes, and in the West big globe artichokes. Low plants are beets, cabbage, broccoli, cauliflower, celery, chives, cucumber, squash, kohlrabi, kale (unless you let it grow tall), leeks, parsnips, rutabagas, spinach, and turnips. Some of the newer varieties of bush squash, if well nourished, get to be three feet tall and four feet across. We had zucchini of that size a year or so ago.

for the birds

The planning we enjoy as much as any is that which we do to attract the birds to frequent our garden. We never claim that birds will eradicate insect pests, but we do rely on our phoebes, swallows, and other insect-eating birds to help keep things under control. We also rely on our brown thrashers to help eat up the poison ivy berries, and the bluebirds, juncos, purple finches, catbirds, starlings, and sparrows to help them out. Ragweed is eaten by redwing blackbirds, starlings, bobolinks, cardinals, all the different sparrows, the tufted titmouse, and the common redpoll. The sparrows also eat crabgrass seeds, and so do the mourning doves and bobwhites.

Birds like tart, wild berries. In fact, they will prefer them to any cultivated berries you have. Therefore plant such bushes and trees as red and black chokecherry, barberry, hackberry, honeysuckle, mulberry, bayberry, staghorn sumac, mountain ash, buckthorn, Virginia creeper, and the various viburnums. They also like the mild blueberries, partridge berry, and huckleberry as well as Russian olive and evergreens like cedar, arborvitae, hemlock, and the various pines. Many of these provide nesting sites and the shelter that nervous birds also seek, especially near where they are feeding. Have a bird bath, with a slight movement of water if you can manage it, and provide nice nesting materials like string and hair.

know when your vegetables will mature

Plants taking a long time to mature include Brussels sprouts (100–120 days), celery (110–130), leeks (140–160), onions from seed (100–130), parsnips (100–140), and rutabagas (100–140.) Those taking a good long time, also, are: onion plants (90–120), pole lima beans (90–115), eggplant (85–100), winter squash (90–125), pumpkins (75–100), tomatoes (80–100), and sweet corn (65–100). Some of the newest varieties of various vegetables are bred for faster maturing, so keep watch, and try out some of them as a supplement to the standard varieties you will choose. Some vegetables can be either early-maturing or late-maturing, depending on the variety.

Planning beans involves many considerations, but a good sequence of supply should be one of the top ones. Include both bush and pole beans, for they mature at 60 days and 70 days, respectively—and last up to 20 days as a crop, if you keep picking them. Also space your plantings, and you'll be able to stretch the season right through until frost. Keep plants far enough apart, especially in dry climates, so they can get nutrients well—2 to 3 feet apart, depending on your climate. The distance between rows depends on your mulching and cultivating program—3 or 4 feet apart if you are going to take a cultivator down the rows, but nearer if you are going to mulch and weed (if you have any) by hand.

The especially quick-maturing crops are radishes, spinach, lettuce, and cress, as well as the first returns you take from the little bulbs for onions called onion sets. You can begin to gather young carrots, beet tops and very small beets, young peas, and cabbage leaves after about 40 to 50 days, somewhat depending on which zone you live in. In Vermont we talk about peas for the 4th of July. Since we often have cold, wet Mays and do not get the garden planted until pretty nearly Memorial Day, this is likely to be a slight exaggeration for some. Even so, peas are ready for nearly everyone in this area by the next week. (Peas grow best when it is cool, so the earlier you can plant them, the better.)

Don't worry if all your seeds do not come up at once. They take different lengths of time to germinate. Radishes are the fastest, so if you want to keep track of where you plant slowly germinating seeds like carrots or some of the herbs like marjoram or parsley, mix the seeds, and pull the radishes when ready to eat. This practice means that when you are making your list for ordering, you should add an extra amount for radishes.

maturing time . . .

succession crops

Since it is sensible to have succession crops, that is, crops you keep planting several times, to stretch the seasons, be sure you know at what date in your zone you must stop planting in order to have the crop mature before it freezes up. Some can bear more cold than others, however. Plants that you can plant up to two to four weeks before the first freeze in your area might include such hardy vegetables as beets, carrots, cauliflower, celery, chard, mustard, parsnips, and radishes—even though the carrots, for instance, won't be so good as those planted five to six weeks before. But carrots winter very well, and you should grow plenty.

For succession planting, it is well to know which you have to allow for as occupiers of the ground for the whole season, and those which mature fast enough to harvest and make space for another planting. Some of those you can harvest early are peas, especially the early bush peas, lettuce, bush beans, spinach, early mustard, cress, beets and beet greens, young carrots, onions from sets, and radishes, the fastest of all. The ones that will have to stay in the ground a long time are broccoli, celery, winter pumpkins and winter squash, leeks, potatoes, parsnips (over the winter, too), salsify, and New Zealand spinach. Mix these in with the protection plants like marigolds which you want to stay in the ground all summer, too. And be sure when you replant that you do not interfere with the pattern of harmonious plantings you established in the beginning.

Typical successions you might consider would be: early beets and turnips, followed by collards; early cabbage followed by late beets; radishes and spinach followed by bush beans or all-season cabbage. Early kohlrabi can be followed by lettuce, leaf lettuce by eggplant, early dwarf peas by sweet corn, peas followed by carrots, beets, cucumbers, squash, or lettuce. These successions would do best where the climate is moderate; in colder climates I'd hesitate about planting cucumbers or sweet corn in mid-July. And though you could do it farther south, I'd never harvest potatoes and then plant tomatoes afterward. As a matter of fact, I'd keep

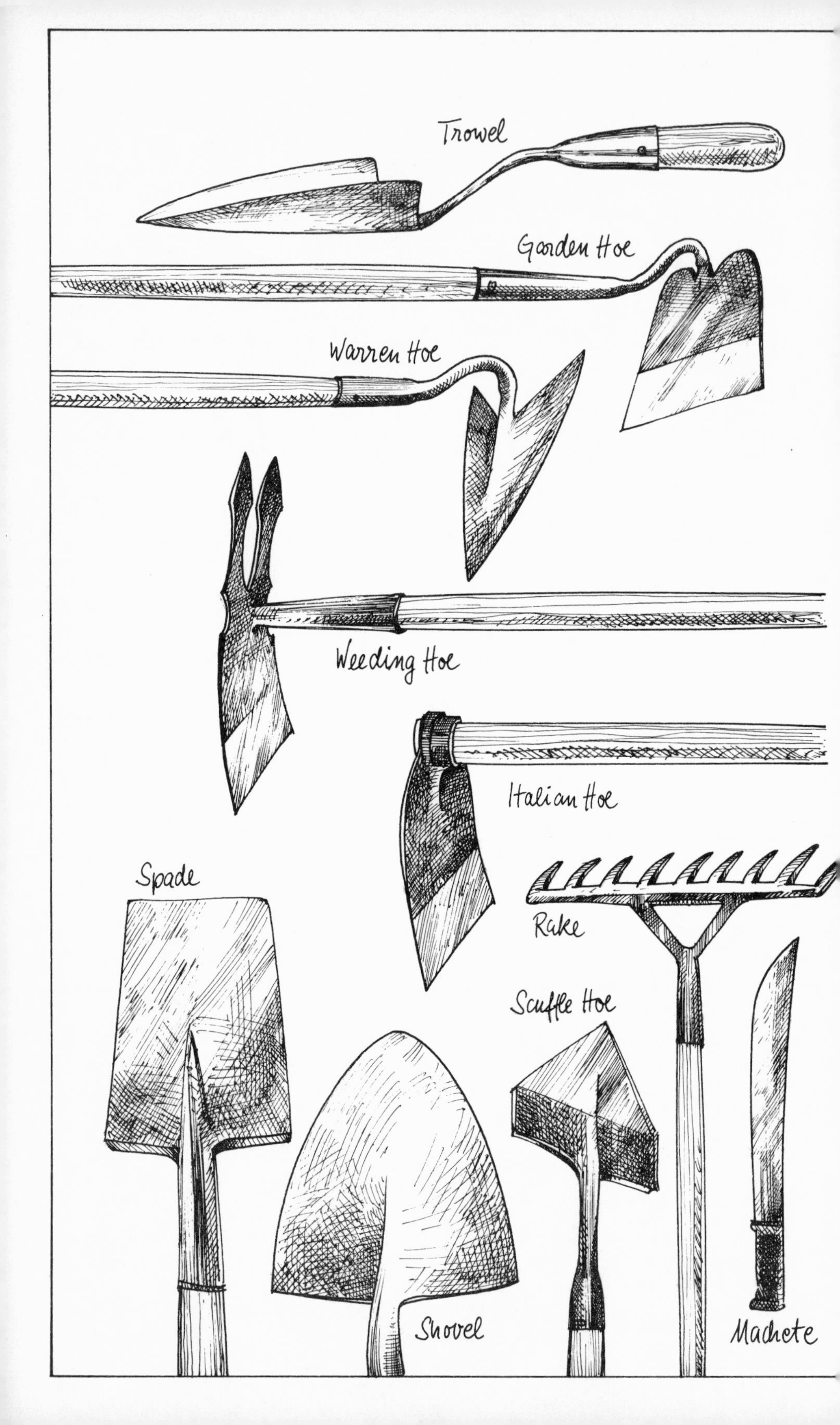
Trowel
Garden Hoe
Warren Hoe
Weeding Hoe
Italian Hoe
Spade
Rake
Scuffle Hoe
Shovel
Machete

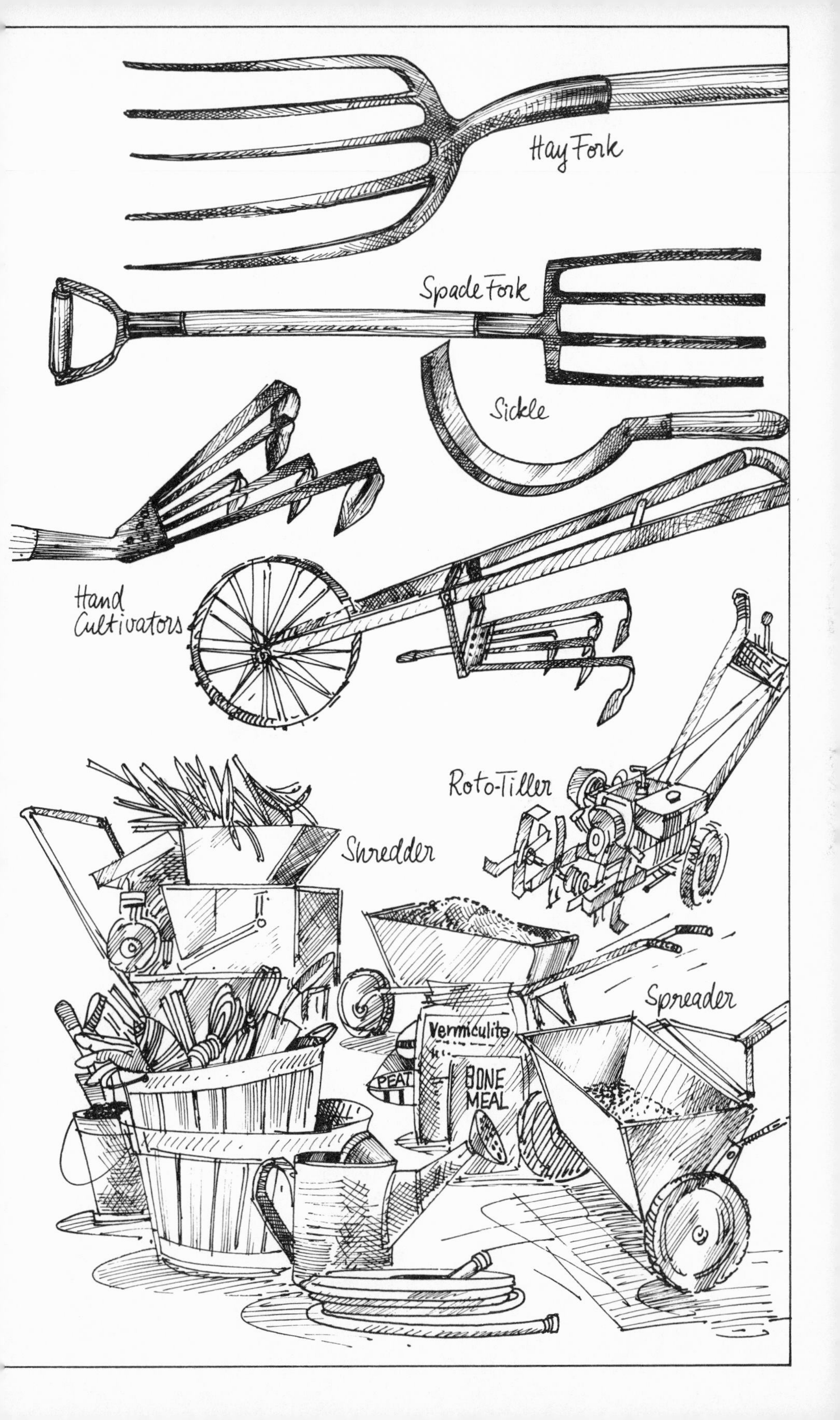
Hay Fork
Spade Fork
Sickle
Hand Cultivators
Roto-Tiller
Shredder
Spreader
Vermiculite
PEAT
BONE MEAL

potatoes and tomatoes away from each other altogether, for pest reasons.

the tools you'll work with

In the toolshed we keep the basic tools we use—a spade fork, with wide tines, for spading in compost and mulch or other organic matter; a hay fork for picking up hay, weeds, pulled-up pea vines; two spades for digging holes, moving earth, plants, sods; two or three trowels of different weights and handgrips; a small hand cultivator which we push through the garden ourselves because as yet we have no Rototiller or small rotary turnplow; one long and one small clawlike cultivator which scratches up the soil and lets some air in when you use it to help take out weeds. There are also several hoes of different shapes—one square-ended, one pointed, and one oblong. And several rakes, including big ones for autumn leaves. We also have baskets, a box or so, two wheelbarrows, two rolls of nylon rope which are attached to short stakes at the ends (and they double for dog-run ropes at times). One of the baskets contains clippers, grass shears, gloves, and at least one trowel, handy to take out for small chores. Bags of bone meal, cocoa shells, peat moss, granite dust, and other such supplies are stored here. One of the best aids is a bucket of oily sand—which will store tools for you, or just be handy to slide a tool up and down in to clean it and oil it before storage on the wall hung between two nails for a prop. Up on the wall also are the scythe, sickle, pruners, and two of the rakes.

Clay pots, Ferto-pots, a basket of dried manure, a bag of vermiculite, a bag of sphagnum moss (the sterile version of peat moss), and a pail of good topsoil are on the shelf for potting or transplanting. Though we do not have it yet, this area will some day get a Vitalite fluorescent lamp over it for between-season starting of delicate seedlings.

At the far end, beyond the big equipment like the lawn mower, is a root cellar, which is not a cellar but a room, insulated and with sturdy shelves for the storage of squash, potatoes, and heavy root vegetables.

We also have a triple-length hose, several watering cans, and one sprayer, which we use for atomizing garlic spray. There is a fertilizer spreader for bone meal.

Of all these tools, Sam Ogden says the hoe is the one that is most worth having—he cuts his down so it is a pointed hoe like a Warren hoe. I think the one my husband likes best is the hand cultivator because he can do so much work with it and get such good results from weeding, aerating, and neatening all at once. For weeding and poking around I like the medium-sized trowel best, though at times, I've noticed, it is just a sharp old kitchen paring knife that seems much the most satisfactory for weeding and aerating.

You will find that you get to have a favorite too. In addition to those we have, you may want to include a shredder, a rotary turnplow or Rototiller, and if you are still in the process of deciding about a lawn mower, you will perhaps choose the rotary lawn mower because it is very good, also, for shredding materials for mulch and the compost heap. Other aids for weeding include a double-disk implement which you push along just under the surface of the soil, and several that slash like a machete or sickle; scuffle hoes that chop and dig; and nursery hoes with sharp points as well as the square digging blade.

Winter is a good time to travel around to various hardware stores and agricultural supply places to inspect and choose what you want. It is also a good time to get things ready like the jar you will need for collecting and annihilating Japanese beetles (and the gasoline, kerosene, or paint thinner), and the boards, sections of plastic mulching strips, lengths of plastic hose, or whatever you decide to use for trapping earwigs, slugs, and other night-working insects. If you are going to try out plastic mulching, order black plastic in big pieces. It will serve many needs, including covering for your composting materials. You can also store up tanglefoot or a molasses and fiber mess to catch bugs going up tree trunks. You can accumulate mulches of some sorts—ground bark, sawdust, and peanut hulls, if you can get them.

The thrifty and canny organic gardener will train himself to make, store, and save things for later use.

For example, you can plan to make birdhouses and feeders. (See Appendix.) You can make various new kinds of traps and enticements for bothersome insects. You can find new tools, hoard aluminum foil for an insect-fooling mulch, and save your torn nylons or cut-off legs of panty hose for filters and for nice, soft ties for plant support. You can also gather materials for making a frog pond, some beehives, and the poles or supports you may want for tying up the plants like tomatoes and peas which may need it.

insect traps

Some of the insect traps worth considering are:

A screen trap over a pan of sweetened water. This trap is usually made about 24 inches high and 12 or 14 inches square, like a screen box, but with one end open and a screen cone inserted, wide at the entrance, small where the insects would emerge in search of the sweet water. The sweetener might be molasses, sugar and fruit wastes, aromatic oils like oil of sassafras, or some protein like powdered egg or powdered yeast. The amazing number of pests attracted to such bait includes cucumber beetles, corn borers, corn ear worm moths, and moths of army worms, cutworms, and tomato horn worms. It will also bring in cabbage loopers and wire worms. If you put out such a contraption on a summer's night when these creatures are on the move, you'll be quite gratified with the results.

For hopping insects like grasshoppers you can rig up a long trough, with a backboard up to three feet high. It might as well be high enough to knock back the creatures when they jump up to escape. What they are knocked back into is the trough, which you can fill with kerosene, or water with a film of kerosene on top. This trap can be any length that is convenient to pull along a row—probably not more than four feet. It can also be provided with two runners and a rope handle to make it easier to move. If you want, you can coat the backboard and sides with some sticky stuff like tanglefoot, and plan to pick up some extra pests with that. This sticky stuff should probably not be put on until you get ready to use the device during the summer. When you buy the sticky stuff, get enough so you can smear it on pieces of screen, cardboard, or floppy, heavy plastic, which you can hang in the wind. This too will attract and trap insects. But remember this warning: examine what you catch each morning. If the catch contains your best friends among the insects—ladybugs, lacewings, or trichogramma wasps—stop using that kind of all-night sticky trap and use something else.

The same holds true for the screen box mentioned above, and for the various kinds of black light lamps now coming into popularity.

One effective version of this lamp trap which you can make yourself is a trap devised at the Purdue University agricultural station. It has a fan, put at the top over a screen cone leading to a collecting can. Over or near the opening is an arrangement of one or two ultraviolet or near-ultraviolet tube lights, available for about $1.50 or $2.00. These can be run by batteries or on a long extension cord.

Traps

insect traps . . . There is also plain old-fashioned tanglefoot. All you have to have is a supply of this sticky material, or a roofing tar substitute, and all you have to do is paint it in a circle around any trees, like apple trees or cherries, which might be infected by worms that crawl up the trunk. It traps them all.

If you can get hold of some bamboo lengths, or old rhubarb stalks, or plastic hose, you can cut one-foot pieces and tie half a dozen together, preparatory to setting out earwig traps. These pests like dark, moist places, and will get into your lettuce and broccoli as well as into your keyholes and behind any loose boards you have around the place. They will also go into a crumpled paper or rag. When you leave these dark lengths around among your plants for several nights and days, you'll find that a good number of the earwigs will want to get into them. When they are collected, dump them out in your frog pond, or put near your toads' holes. One of my neighbors who keeps frogs says that it seems no time at all after his discarded earwigs hit the water before the frogs in his pond have them all gobbled up. In fact, such a little frog pond with such an insect-disposal system is a lot more attractive to have than just a can with kerosene in it.

Winter is a good time to plan that pond and also, of course, if you have room for it, to plan a farm pond for fire protection and fish. Obviously fish like to be fed insects, too. And I have heard that pound for pound you can get more food value from enriching a farm pond than you can from an equal expenditure of enrichment on the land to grow crops. I believe that this was calculated on the basis of commercial fertilizer, but the principle would be the same whatever you used, I should think.

organic sprays

Most of the organic sprays you will make will be done at the time of need, in the summertime. But it is possible to make some in the winter, and freeze them and store them until the time you want to melt and use them.

A very useful spray, suggested by Rose Houskeeper of the Garden Club of the Oranges, can be made by grinding in a blender 1 medium-sized onion, 1 teaspoon of very hot red pepper, 3 cloves of garlic, and a quart of water. She suggests that you strain this juice through a nylon stocking right into your sprayer, but for winter storage, strain it into a large cottage cheese carton or whatever you are going to freeze it in.

chapter four

compost-ing for the cycle's sake

If there is one thing as an organic gardener you are really devoted to, it is your compost heaps—and especially your participation in the methods of composting. You take part in the most basic cycles of nature, accumulating richness, dispensing it, making life out of leftovers and healthiness out of materials that are going to rot. Ever since Sir Albert Howard urged farmers and gardeners to return to the soil all that we take from it, the organic gardeners who follow him have put faith in the composting methods he devised for doing just that. You make it yourself because you can't buy it.

While scavenging, saving, or hoarding things like potato peels or apple skins that are too good to throw out, you know that in the processes of nature there is no throwing away. All is used for one system or another. We talk about things deteriorating, but also, every time, there is a building up of something new. Any plum or peach that is going to rot is providing food for the bacteria or molds that therefore thrive. And they, in turn, work on the stone to free the seed. The main base for the penicillin mold so widely used during World War II came from a cantaloupe found rotting in a grocery store. And all modern penicillin comes from mold. You learn to put every organic material you can onto your heap, and you go on the hunt for new sources to pile up new stores of enrichment. If you are not a scavenger, and take no pleasure in accumulating materials this way, you can buy many of the ingredients you need.

Your scientific knowledge helps you, too. You learn that though tons of nitrates in this country are washed down the rivers and into the sea, the store of nitrogen needed by plants can be kept from erosion and built up to transfer to the soil. Then you get twice the urge to create that store, and promote that transfer. Since good soil health depends on organic matter to keep it rich and loose, you want to provide your garden with a big, rich source of it. You discover that soil gets pale and deteriorates over the long stretch when chemical fertilizers are substituted for the organic fertilizers found in nature, and then you are even more convinced that making compost and spreading it on the garden are absolute essentials for sustained, productive gardening. You also know that you are not overfertilizing and flushing excess nitrates and phosphates into the environment.

where to put the heap

First you find a good position for a heap, or two, or three. According to the thirty-year experience of Samuel R. Ogden, three bins are best (though now in his old age he says two will do). Ogden, who is a canny and successful Vermont organic gardener, came to this conviction not only from his master, Ehrenfried Pfeiffer, but also from his own well-thought-out labor-saving habits. But in the suburbs or in a city plot one heap is enough. In fact, many people settle for one in any location. Heaps have to be aged and turned, and with three you have one coming to perfection every year. For this three-year cycle, you must pick a site that is long enough for three heaps. That might be twenty-four feet long, for the usual dimension for each bin is 5 feet by 5 feet by 8 feet long.

The site should be flat enough so that the nutrients accumulating will not leach away, and will ripen in a well-moistened condition—up to 50 percent water, but aired and not soggy. To hold moisture you can try to keep the heap somewhat concave on top. It should be far enough away from trees so that the roots of the trees will not come up into your piles and help themselves to all the goodness. It should be near enough to the garden and near enough a driveway or access of some sort so that any heavy materials you bring to add to the pile will be fairly easy to handle. A site within the stretch of your hose is advisable too.

Since neighbors are what they are, it is a good idea to hide your compost heap. They will have all sorts of myths and misconceptions about what a compost heap will do, and may even look on your carefully constructed, scientifically planned source of soil conditioning as a garbage pile. First they'll ask you whether you will get rats. Then snakes. Then dogs and cats. If you explain that you put dirt over the garbage, and leaves and grass clippings, and if you add that you bury the manure with wood chips, or sawdust or more dirt, they may relax. If you also explain how fast the bacterial action is when the pile heats up, and the conversion of scraps, excrements, germs, weed seeds, and wood chips thereafter to clean, black humus, maybe you can satisfy their questions or even arouse their admiration and wonder.

the heap . . .

You might add that if you do attract dogs and cats, they will obviously control the rats, in case you might have attracted rats. The chances are very slim, however; I've not known anyone who attracted rodents any larger than mice—and those, in fact, moved in under the warm and sufficient haven of the mulch, not the compost.

Those lucky enough to have an owl or an opossum have found he will help to control rodents, too. In any case, the neater and more controlled you make your compost heap seem, and the more you disguise it, the easier your life will be. You might even try growing some morning glories or a kudzu vine to bedeck the bin or fence.

sheet composting

Some people, if they find that the compost heap continues to bother others, simply move their composting activities to the cellar, the garage, or a closet and use boxes or plastic bags as described in Chapter 2. Others, if the crops are up high enough to disguise what is happening, toss the stuff directly on their garden soil. Ruth Stout, author of *How to Have a Green Thumb Without an Aching Back* and other books on organic gardening, recommends a good, deep hay mulch on your garden, so you can just slip your garbage out of sight underneath it. We use the euphemism "sheet composting" for these methods.

constructing the bin

If you have room in your yard, and feel like building something, a slatted bin with removable slats along one side is a very good-looking and handy construction to put up. When you want to take out the finished compost, you simply pull out the slats and start filling up your wheelbarrow with the black, fine-grained material. A very easy substitute for this sort of bin is one made of snow fence, with four stakes at the corners. It is even easier to put up a single circular fence form, and then have only one key stake to worry about. We have used this kind, but find that a few extra stakes will keep the fence from sagging, especially after it is exposed to heavy snows in the winter. The most rot-resistant woods to use are cedar, cypress, and locust. (A few gardeners still consider redwood, but most of the organic gardeners I know

Compost Bins of simple Construction

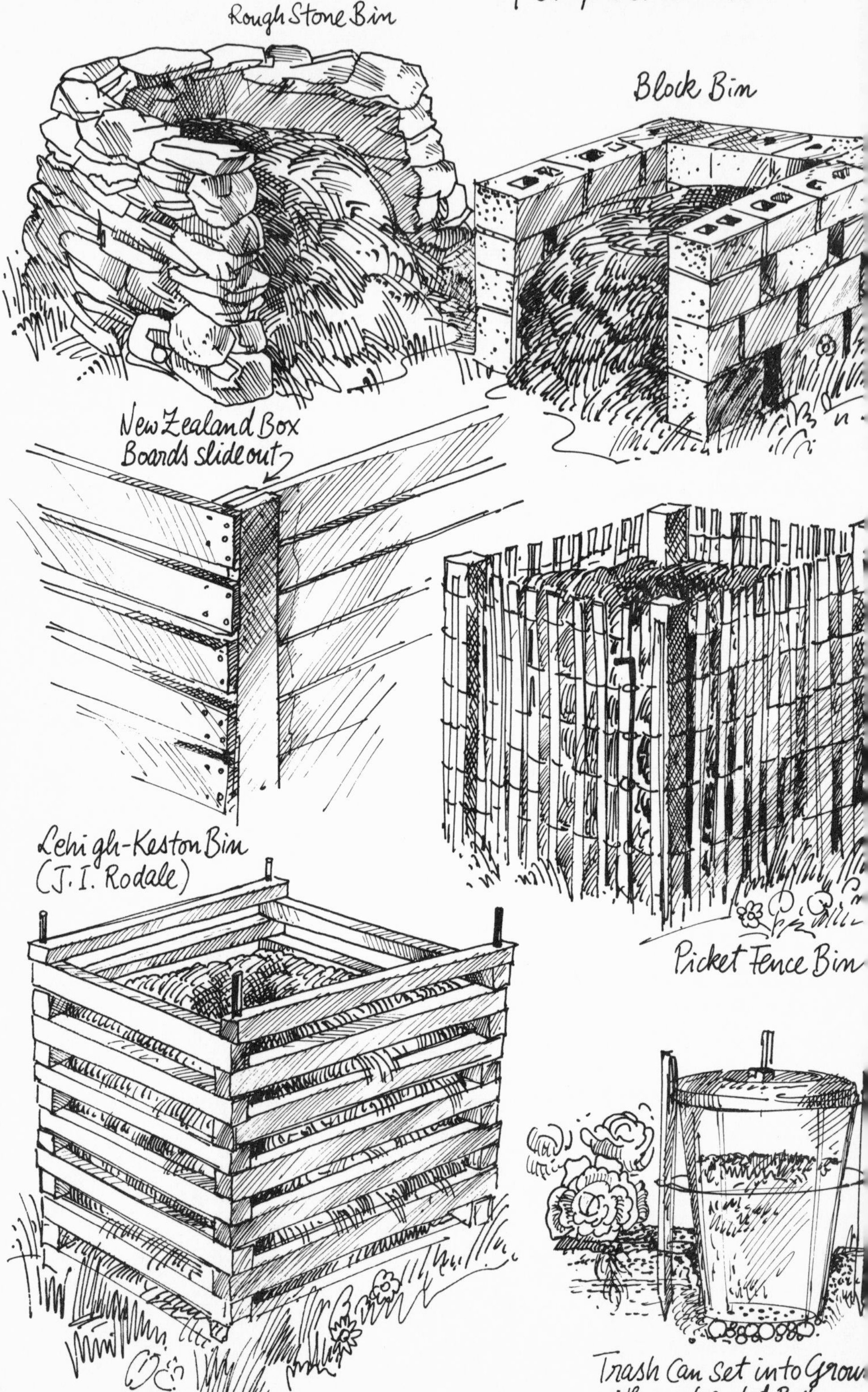

Rough Stone Bin

Block Bin

New Zealand Box
Boards slide out

Picket Fence Bin

Lehigh-Keston Bin
(J. I. Rodale)

Trash Can set into Grou[nd]
with perforated Bottom

wouldn't dream of it, for they are ardent conservationists and belong to the Save-the-Redwoods Federation.) Use an epoxy paint to discourage fungi. Do not use creosote.

the bin . . .

The Scott Nearings make a bin with plain thin logs. Both in their Vermont organic farming and in more recent years on their organic farm in Maine they have had excellent results from a bin laid up log-cabin style with four new poles set in place whenever the composting materials reached a height where more support was needed. Scott Nearing always keeps the heaps squared instead of piled for that good dip in the top to hold water.

Not so easy to handle as logs or boards, but very easy to come by, are concrete blocks, and many people now use them. If you build up the bin with alternating air spaces, blocks make an airy and durable frame. You can pull out any section you want when you are ready to turn your compost over or take it out for use. Some people don't like the way they look, but these blocks can be disguised with paint or vines if you want to hide them. You can also have the first row or two below the surface of the earth and begin your composting in the hole you make. For those who want to use natural materials, stones can be stacked up to look like the stone beehive cabins at the western tip of the Dingle peninsula in Ireland. Other compost makers use bricks, and if they are an old mellow pink, they can be very attractive. Always leave easy access for turning the heap, or make the structure three-sided in the first place. You'll have more trouble keeping the dip in the top for water if you do that. But as most gardeners are well aware: you can't have everything.

air needed as well as water

Aside from water for your heap, you have to have air, which is absolutely necessary for the aerobic (oxygen-using) bacteria who do your composting work for you. You can fork the pile over or insert several two-inch hollow pipes into the center, put in from the top or from the sides. People believe that by using such pipes, they do not need to turn the pile. If it cools off, however, you'd better get busy at once to let more air in. After once heating up, if the pile cools off before the compost is made, it will take a long time for the bacteria to bring it back up to the proper heat of 150 degrees. Quickly add nitrogen (for instance, blood meal) to help them along. To keep track, tie an oven thermometer on a string or thin wire and insert it into your heap. When in doubt, pull the *air, water . . .* thermometer out to see how your pile is doing. After a while dig in in several places, and if the materials have turned to a rich, soft blackness, and are again cool, the compost is ready to use. If you live in a very rainy climate and need especially good drainage, you may have to build a bin that is totally above ground. But construct it with a coarse screen bottom at least a foot above the soil surface. The dark, fine compost that forms at the bottom simply falls out ready for spreading. If the stuff falls out too soon, fit in five or six layers of newspaper to hold it in at first. They'll rot and shake loose at just about the time you want the compost to come down.

Organic gardeners are constantly coming up with new ideas. Someone invented biodegradable walls, which can be made of many piled-up sods. The rest of us like the idea of this sort of bin, with self-destruction built right into it, for the whole thing will eventually break down as compost, and you need have no unhappy anticipation of shabby blocks or a ragged, leftover wooden structure to bother you when it gets worn out.

One farmer I know uses bales of spoiled hay for the walls of his compost piles. Some of the bales can be moved easily when it is time to turn the heap. They are biodegradable and are a very handy source of hay mulch if you need to save moisture or protect little plants that have just become established.

Most inconspicuous of all are sturdily lined compost pits on a hill with the side facing downhill used for the opening. They can be put right next to the garden or, if you want a handy place for weeds, right in the middle of it. Keep them small enough to work with, not deeper than four feet. The proper sizes for pits and heaps were discovered by Sir Albert Howard, when he first began composting. Later he had his helpers make very long heaps; but short or long, the intent was to keep the dimensions convenient for turning. The bottom is always bare ground so the earthworms and bacteria can get in. Pits are very appealing to those who like to bury things. And they also save heat. With well-shredded materials and careful management, compost can be made in a pit in quite a short time, and can be insulated for winter warmth.

earthworms

With a lined pit you are not likely to get as many earthworms as you would in your compost heap, where they *earthworms . . .*

are readily attracted to the rich organic matter. Many organic gardeners even buy worms to insert in a layer near the bottom. Earthworms condition the materials and help in making humus, both in the compost heap and in the garden. Through their castings and movements in the soil, they open it up to create the kind of structure the gardener wants for his land. It is as though earthworms ate their way through the compost pile or the soil. They suck both minerals and organic matter into their mouths as they move through the earth. In addition to eating decaying organic materials, they eat live plant seeds, larvae, eggs, and parasite cysts. This food is ground up in the pharynx, then neutralized by calcium carbonate secreted from glands in the esophagus, before going to the crop and gizzard. In the gizzard the food is thoroughly pulverized by the small stones and mineral particles ingested by the worm, before it goes to the intestine, where it is digested by enzymes. The castings, therefore, are a good deal nearer the state desirable for plant use.

When you get ready to use compost on the garden, save the worms from your pile by heaping the materials on a tarpaulin or big plastic sheet in the warm sun. In a few hours, when they react to the heat of the sun penetrating to them, the worms will cluster in the middle of the pile where it is coolest, and will be easy to pick up all together and move.

One man I heard of collects his worms that way, puts his compost on the garden in the fall, and moves his worms to a special insulated bin he builds so he can keep them warm and active in the winter. He adds nutrients to the inside, manure, hay, and other materials stacked up to the top, and insulation materials in an encircling outside bin two feet thick. (One foot is thick enough in climates less chilly than ours in Vermont.) This way he keeps his worms busy, has new compost ready in the spring, puts it all on a tarp again, and shifts the outside materials to the inside to start once more. A splendid triple-purpose system.

winter methods

Other methods for winter include covering your heap with dirt, sawdust, leaves, a tarp, or hay—all to keep out the wind and prevent the escape of nitrogen. Some very careful composters believe in covers all year round. I find that the most recently added vegetable wastes serve that purpose, or the most recent layer of dirt. The tarp, of course, will fend off animal visitors. For leaves get oak leaves if you can, because *winter methods . . .*

they are highest in nutrient values. If you have none of your own, scavenge from the town park or road crews who go around gathering up leaves in the fall. Since you will probably be saving them a trip to the landfill, they will be glad enough to let you have them. Sometimes, if you live on a convenient route, they will even deliver them to your yard. If you can't get oak, get maple, ash, or tulip tree leaves.

Before adding the cover for winter, test the pH of the heap to see if the materials are too acid (below pH 5.5) and need lime. Sprinkle in enough to raise the pH to about neutral (2 cups of limestone or 5 of wood ashes). You can also Rototill all your compost into the garden in the fall, and not have any heaps to winter over.

Plastic bags can be used in seasons other than winter. You can make your own, but I prefer the big dark green garbage bags now available in markets. Use two or three for extra thickness, and because the outside one may split when you take it out into the hot sun. Since the bags will be tied up, you can stuff in weeds, grass clippings, all kinds of green stuff, handfuls of earth, manure, bone meal, and whatever else you want. Leave the bags out in the warm weather, and you can expect a rapid and complete composting as they heat up in the sun. Don't leave them too long or the bags will split. When it rains or at night, they can be hung on hooks in the garage, or down cellar, or wherever you want. A friend of mine uses gerbil cage cleanings, bird cage cleanings, vegetable garbage including some of the old, worn-out lettuce she begs from the chain stores, and handfuls of grass she pulls off as she goes by the end of the yard. In the summertime, with four bags going, the first is just about ready to use when she starts filling the fourth bag. This is good speedy composting.

Though there are periods when the smell can be rather awful as you open the bag to add new stuff, the ardent composter will still like this anaerobic method, for there is no worry about keeping the heap well watered, well drained, and well aired, and no need to turn the heap, for the anaerobic bacteria don't need air. There is no worry about leaching—either up to the air because of windiness and drying out, or down into the ground because of rain. All is contained, and working in a small well-controlled place. In fact, the more you think of it, the more you come to see that except for the smell there is no good reason not to compost by this method. Well, there is one good reason, and it is related to the basic reason a gardener likes his occupation so much. That is that he can be out of doors, working with things in the open, and enjoying the four elements of earth, water, warmth, and air. Everything about the anaerobic method seems tied in and confined in comparison.

winter methods . . .

You may have seen ads about adding bacterial activators to compost-making materials. I cannot see that such activators are needed, for the bacteria come naturally from the air, the soil, the plant materials, and especially the manure used, which obviously has bacteria already. Under favorable conditions the bacteria will undergo enormous population growths, and you will have as many as you would if you had bought them. The Bio-Dynamic group of gardeners uses an herb supplement as an activator for nutrients. That's different. I'd suggest you save some compost material to use as your own activator or seed compost on the next compost heap you start both for nutrients and for bacteria. If you are just starting, borrow some seed compost, or rely on the manure.

use everything of organic origin to make your compost

Most householders first think of grass clippings, autumn leaves, and garbage as the materials to use in their compost heaps. All are excellent, and at least two out of the three you can use for mulch materials first. In fact, a basic source of composting materials is last year's mulch. Green grasses, especially young blades, are high in nitrogen, so are good for both mulching and compost. You'll find yourself wishing you had three times as many lawn clippings as you do, for they do the lawn a lot of good when left where they fall, they are excellent as mulch to spread directly on the garden, and they do well on the compost heap, especially if you add blood meal along with them, or fresh manure. Spread them around in all three places, or mete them out according to the weather and the current needs. But I think best of all is to use the spring clippings as thinly spread mulch on the garden when the first vegetable seedlings are coming up. The nutrients are good; and the fine grass will not choke the tiny plants as some bigger, coarser mulch might. Do not spread thickly. They'll get hot and slimy if you do.

The leaves from a good big shade tree will provide your garden with a lot of nutrients, up to twice as many as you'd get from an equal weight of manure. In fact, plant scientists have estimated that you can get back from a plant just about the amount in nutrients that went to make it. Perfect cycling would take place if humans never entered *materials for compost . . .*

the cycle to take off crops of plants, and meat or milk from animals that eat the plants. In dollar terms, considering the cost of fertilizers, peat moss, and mulching materials, the leaves from a big tree might add up to the equivalent of nearly twenty dollars. For best results they should be run over with a rotary lawn mower (or by cars on the driveway), or put through a shredder before composting them. If they are spread right on the garden and allowed to rest there over the winter, whether or not shredded, they can be Rototilled or plowed in in the spring. Rototilling or digging in is better. By midsummer most will have disappeared except, perhaps, in a few places where there has been a heavy bunching up of unshredded leaves. Don't let that happen. Your plants will get nitrogen starvation and be sickly and stunted.

Vegetable garbage is an indispensable ingredient of a compost pile for several reasons, and be sure to include coffee grounds and tea leaves. Such garbage is full of a wide variety of nutrients and growth factors, which supply the bacteria, fungi, actinomycetes, earthworms, and later the plants with both their major needs and the trace elements. In addition, you get the satisfaction of putting them to good use in the composting process that transforms these wastes into something clean and wholesome. To take part in cleaning up and enhancing the life of the world around us is exactly what the organic gardener aims for.

Residues from the kitchen can be shredded before adding them to the compost pile. I put a lot of my vegetable garbage through the blender before adding it to the pail. I also grind up eggshells and all such things as melon rinds or orange and grapefruit peels that take months to degrade if left in large pieces. The water you use for blending is needed for the compost heap anyway, and the many small surfaces are good for the bacteria to work on. The earthworms are not fussy, but some gardeners think they like a finer-textured pile to work in, too. Earthworms particularly like sawdust, ground bark, and wood chips, which are excellent for composting; so do fungi. Any acidity in the heap is only temporary. Don't worry about it.

Newspapers are all right for composting, too, especially if shredded. For $110 you can get at an office supply house (like Goldsmith's in New York) an electric wastebasket that will shred paper for you. It ought to be adequate for composting second-class mail, too!

All weeds can be added, as well as plants culled when you thin the rows, and those left over after a crop has *materials for compost . . .*

been harvested from them. This certainly applies to cucumbers, carrots, cabbages, and lettuces and the plants of all their families, unless you have cabbage club root (see p. 135) on the cabbage family plants. (Burn them and use only the ash.)

The legumes that have finished bearing in your garden, such as peas and beans, can be added to the heap or pulled up and heeled in right there on the spot. You can improve them by running over them with a small Rototiller to chop them up. This makes their plentiful nitrogen available quickly for your compost heap or garden with as little loss as possible.

Incinerator and wood ashes are very fine to use, but they leach easily, so it is better to put them right on the garden. They are also useful to put on the cabbage worm if it decides to come in.

Do not worry about disease germs and weed seeds. The heat generated by the bacterial action during composting kills off all but a few fungus spores. The only seed to come through alive, in fact, is the tomato seed. I have friends who never buy tomato seeds or plants; they just wait to see what will come up. If cherry tomatoes get started, they will pop up all over your garden for several years. Whenever it is completed, spread compost on your gardens, lawns, around trees and shrubs as needed. In the fall, half-completed compost can be put out to weather over the winter, under the snow—if you live in the right climate.

getting the major nutrients

nitrogen

Convenient and reliable sources of nitrogen for your compost pile are: blood meal, or dried blood, which has a nitrogen content of 12 to 15 percent; cottonseed meal, with 7 percent; feathers, more than 15 percent; hair, about 14 percent; slaughterhouse refuse, called tankage, with 6 percent; bone meal, 5 percent; activated sludge, 5 percent. If you can get hold of some of the following, the nitrogen content is very favorable: ground leather, 11 percent; dried shrimp heads, nearly 8 percent; meal or dust from hoof and horn of cattle, 12 percent; waste from wool or felt hats, 3 to 4 percent; ground-up lobster shells, nearly 5 percent; fish scraps (as we learned in school when studying the Indians) yield 6 to 8 percent, and the very special King crab, 10 percent. Fish products can often be found in garden stores as fish meal, or as fish emulsion, but percentages of these are lower.

Of the manures there is quite a difference between the wet and dried nitrogen percentages. Dried manures pro-

nitrogen . . . vide 3 to 4 percent more nitrogen. Here is an estimate for wet manures. I've seen others even lower.

Cattle manure	.29%
Duck	1.12%
Hen	1.63%
Horse	.44%
Pigeon	4.19%
Dog	1.97%

Nitrogen is needed in every cell, for good greenness, and for rapid new growth.

phosphorus

Higher yields of the essential element phosphorus can be had by using some of the following on your compost heap. Residues from a sugar factory, if you live near one, will give 8.33 percent; ash of lemon skins if you incinerate such a thing; 6.3 percent; ash of cucumber skins, 11.28 percent, and dog manure, 9.95 percent. Other good sources for phosphorus are ground cocoa shell, 1.49 percent; common crab, 3.6 percent; Milorganite, sludge from a sewage treatment plant, 3 percent; basic slag from a steel mill, 8 percent; cottonseed meal, 2.5 percent; cottonseed-hull ash, 8.70 percent; castor-bean pomace, 2.25 percent; apple skins (ash), 3.08 percent, and bone meal way up to 21 percent or steamed bone meal at 27 percent. Burned, ground bone gives up to 34.70 percent phosphoric acid, and that's about the highest compost element you can find—unless it's waste from a paint mill at 39.50 percent. Fish scraps from red snapper can yield 13 percent, and municipal incinerator ash, if you can get it, 5.15 percent. The ash of raw potato skins gives 5.18 percent, sardine scrap, 7.11 percent, and siftings from oyster shell mounds, if you live on the Chesapeake Bay or certain coastal areas of the Gulf of Mexico, 10.38 percent. What you are aiming for with a good supply of phosphorus is a good root growth, sturdy growth of stems, good flowering and fruiting, and—so it is reported—an increased vitamin content in your plants.

potassium

Potassium, or potash, is the third major need in the supply of nutrients you want for your garden. It strengthens plants, makes them disease-resistant, is needed in cell development and cell division, and helps plants use and control the nitrogen they receive. Though it is difficult for plants to use potassium in all forms, incinerator or fly ash is helpful, as well as powders and dusts which contain potassium such as greensand and granite dust, which have 7 percent and 5 percent, and also various vegetable residues. If you can get it, fly ash has 12 percent. A whopping 49.40 percent of the ash of banana stalks is potash; and 41.76 percent of the ash of the skins of bananas is potash, also. Ash quantities of others include: apple skins, 11.74 percent; cantaloupe rind, 12.21 percent; corncob, 50 percent; cottonseed hull, 23.93 percent; and fire-pit ashes from smokehouses, 4.96 percent. Milk will yield .18 percent; oak leaves, .15 percent; the ash of pea pods, 9 percent; and wood ashes, 7 percent. (You don't have to burn these things. The figures are to give you an idea of the inherent nutrient values.)

potassium . . .

Wood ashes are the old-time gardener's stand-by for potash. He guessed something about the corncob, too; but he was quite ignorant of the beneficial potash value of the ashes of apple skin, citrus-fruit skin, and potato skins, let alone of pea pods and bananas. Nevertheless, an age-old ingrained knowledge or intuition about the soil and its needs has dictated to many farmers that it was better to put wood ashes back on the land than it was to remove them from the cycles of plant nutrition. Country people also understood the value of sea products. Fields in old-time cultures like those on the Channel Islands off the coasts of Britain and Brittany indicate an age-old knowledge of sea-product fertilizers. On the island of Jersey, I have walked beside plowed fields festooned in January and February with the maroons and cream colors of kelp spread along their furrows for fertilizer. As the season for planting approaches, these decorative fertilizers will be plowed under and the rest of their nutrients released for the plants of the Jerseymen's crops.

Obviously all those exotic sources of nutrients for your compost heap are not going to be available without a great deal of effort, and probably expense. But the figures suggest that you save and scavenge all such things as banana skins or crab shells when you can. If it is nitrogen and phosphoric acid your soil test indicates you need, you can even send away for bat guano, which has 6 percent nitrogen and 9 percent phosphoric acid. That's a pretty good proportion—considering that common sources like alfalfa have only 2.45 percent nitrogen, .50 percent phosphoric acid, and 2.10 percent potash, and cowpeas have only 3.10 percent nitrogen, 1 percent phosphoric acid, and 1.20 percent potash.

When they find their soil is low in phosphorus,

phosphorus . . . organic gardeners apply phosphate rock—preferably pulverized and interspersed with pulverized limestone. For potassium they often get greensand and granite dust. Go to your garden center, grain or hardware store, or organic food shop as possible sources for these materials. If you find you have to send away, see list of addresses in the Appendix. If possible, contact a buying cooperative, and get in touch with your local organic gardening club to see whether arrangements have been made to buy in bulk in your area.

A useful multipurpose supply to have on hand might be: a pile of sludge—free—which you can put in the compost for nitrogen and phosphorus; manure—free—for nitrogen and potassium; bone meal—to buy—good for phosphorus, especially; greensand and granite dust—to buy—good for both phosphorus and potassium; and the following good for all three major nutrients: cottonseed meal, seaweed, fish meal, cocoa shells—expensive—wood ashes and incinerator ashes—free, and especially good for potassium.

If you want to spend a lot of extra money for ground volcanic ash (known as perlite) or expanded mica (known as vermiculite), you can create a wonderful loose texture (and very water absorbent) by the addition of such materials to your compost piles or directly to your soil. Cheaper would be chopped-up corncobs, tobacco stalks, or sugar cane. But remember that mice like these materials, too, especially the sugar cane. Do not merely dump. Mix the additions in thoroughly.

trace minerals

In general, trace minerals for organic gardeners are mainly derived from organic matter itself, or from rock phosphate, lime, granite dust, or sludge. The leaves of big deep-rooted trees, for instance, are a fine source of minerals because of the way the roots burrow way down into the parent rock zone and bring up a rich supply.

Boron is needed, and is deficient in the soils of many parts of the country. Good sources for composting are vetch, sweet clover, and muskmelon leaves as well as granite dust. Copper, deficient in some eastern states, can be derived from spinach, tobacco, dandelions, wood shavings, sawdust, and brome grass. Iron, available from many weeds and vegetable residues, is deficient only in most eastern states and the north central states; it is necessary for animal metabolism and can be derived from several of the basic rocks, vetch, most legumes, and peach stones.

trace minerals . . .

Manganese can be got from carrot tops, red clover, leaf mold—especially if from white oak or hickory—and alfalfa. Zinc, which is deficient both in the South and on the West Coast and is needed by both plants and animals, can be found in ragweed, hickory, poplar, vetch, cornstalks, horsetail and peach tree clippings. Vetch, alfalfa, and rock phosphate are sources of molybdenum, which is deficient in many soils.

Calcium, which comes from lime that is very slow to break down to the size of particles usable by plants, is best when finely ground, preferably small enough to sift through a very fine 100-mesh screen. It is not exactly a fertilizer, but it is valuable to adjust the alkalinity and loosen soil structure. The best source, since it is mixed with other nutrients, is dolomitic limestone. For the compost pile, it is enough to sprinkle a little on every fourth or fifth layer, and you can rely on the leaves you add to the pile for a certain amount. Maple leaves have between 1 and 2 percent; oak leaves nearly 1½ percent, beech nearly 1 percent; and white ash, which is plentiful in the temperate zone in many places, has nearly 2½ percent. If you keep piles of leaves near your compost pile, reach down and pull out the wet lower layers when you plan to add some. Not only are these leaves already partly decomposed, they also are less likely to blow when the wind comes up. One more good source of lime is glass. The ground glass frit now becoming available for gardeners has calcium, sodium, iron, potassium, and sometimes boron in it. Several companies are preparing glass materials that will be soluble enough to make them practicable for farmers and gardeners.

humus and the soil chemistry

One of the best reasons for compost gardening instead of chemical gardening is that both the major and the trace elements are conserved and protected by organic fertilizers, whereas chemical fertilizers might release elements so quickly they leach away. Since such tiny amounts are required by plants, it is difficult not to be excessive in applying the trace elements chemically. Your potato crop, for instance, if given too much boron, would be ruined.

The end product of your composting, humus, has other unusual properties in its fine-grained, dark-to-black organic material. According to Nobel prize winner Selman Waksman, it can probably solubilize nutrients from the insoluble state to make them available to plants. It has dozens *soil chemistry . . .*

of organic compounds in it, with fifteen or so different acids, aldehydes, carbohydrates (carbon-oxygen-hydrogen compounds), and proteins (carbon-oxygen-hydrogen plus either sulfur or nitrogen or phosphorus). One analysis of humus showed: carbon, 44.12 percent; hydrogen, 6 percent; oxygen, 35.16 percent; nitrogen, 8.12 percent; ash, 6.6 percent. The ash contained calcium, phosphorus, boron, zinc, magnesium, manganese, and the other trace elements. Though humus is well known to aid storage of nutrients, and promote the right tempo of release of these nutrients into the soil solution, it also has a mysterious power to control the action of the organic compounds. Sometimes organic compounds stretch out or move around in a swimming motion until they reach the mineral particle, to which they will grab on with a claw-like grip. Then the mineral comes out in the open where plants can use it. Plant scientists now believe the mobility of plant nutrients may be largely due to this curious grabbing (or chelating) ability of organic matter. In many scientific reports on the action of organic materials in the cycles of plant life, there comes this point of *maybe.* It seems to me that every one I've seen has two or three *maybes,* pointing to powers and actions going on which man has not yet discovered. Even a 1 percent possibility of error in chemical manipulation still makes humus and compost 99 percent safer to use on your soil.

Yet there is a host of already discovered life processes which go into the making of compost or humus, in which the biological and chemical changes are made cohorts of microorganisms with the help of enzymes to do the errands.

why we want compost and organic matter

Organic matter will have some other important influences: it turns the brown or gray soil to black; it encourages granulation; lumping and cohesion are reduced; it holds six times its own weight in water; it improves the supply and availability of nutrients. There will be good exchanges and easily replaceable cations; there will be nitrogen, phosphorus, and potassium held in organic forms; and there will be continual, slow extractions of elements from minerals by the more or less acid humus.

The decline of 30 to 40 percent of the organic matter in American soils, say soil experts Harry Buckman and Nyle Brady of Cornell, is serious, and should go no further if we can afford to prevent it. But unfortunately, unless man catches on to what's needed to shift our sense of economy from wastage to salvage pretty soon, he may find that the ideas of what he can or can't afford have snowballed so farcically that he will be saying, "Well, no, I can't afford to breathe any more. It is much too expensive to clean up the air."

scavenging

When you are out on a pickup trip for composting materials, swing around for free manures, free old vegetables, cannery, winery, or dairy wastes, or beauty parlor sweepings. Go to the lumberyard; see whether you can't get the lumberman to give you some free sawdust and some free ground bark. If he demurs, or wants to charge you something, ask him what is causing all that smoke coming from his incinerator. Nine chances out of ten it is from burning sawdust, or bark. You might better have it, for you will recycle it, whereas the lumberman is simply polluting the air.

All such materials that you can save or scavenge are good both for composting that takes months or for quick composting that can be done in a few weeks, with some special tricks. Some gardeners still think a compost must be several years old and fine-grained before it is ready for use; others, who claim equal success with their gardens and equally sizable vegetables and flowers in the results, say that a perfectly adequate compost can be whizzed into shape in fourteen days.

composting in a hurry

To use fast methods you hasten the ripening process by physical, biological, and biochemical means. Since the idea is to get the pile to heat up very rapidly, there must be lots of bacteria at work, lots of oxygen to keep them going, and, of course, all the organic nitrogenous matter that they can use. You make the whole heap at once, layered carefully in alternating materials such as shredded dry leaves, manure, and high-nitrogen green stuff and vegetable garbage. Cottonseed, sludge, blood meal or fish meal will do as well as manure. (When you are adding to a compost of autumn leaves, use twice as much manure, for leaves in the fall are low in nitrogen.) Also see that the organic matter is physically multiplied by giving it many, many surfaces for the bacteria to work on.

Water for fast composting must be kept at half the weight of the pile. Hose it well when you first make the pile, *composting in a hurry . . .*

and let it go for a week or more unless the weather is very hot and dry. In moist, warm autumn weather—one of the best times to make a fast compost, for then you have lots of leaves—one watering may be enough. The next day take the temperature—it should be 150 degrees. The third day, turn it all to get more oxygen in, and see whether it has dried out. At this time, to speed up biochemical change, you add enrichments such as rock phosphate, colloidal phosphate, granite dust, greensand, and small amounts of limestone. Use a shovelful of each to fifty pounds of compost material. Turn often for the next ten days.

For shredding compost material, a rotary mower is the preferred machine. Many of the commercial shredders on the market get all gubbed up when you feed green stuff or gooey stuff into them. Dry twigs, cones, and bark are needed to unstick them, but if your aim is for the fastest composting, the lignins in this material will delay the action, and you'd want to avoid them. For shredding vegetable wastes in the house, use the blender, shears, or the gadget for retrieving stuff that has gone through the disposal unit (see Appendix). A large-sized blender called a Hydramill can be used outdoors. It produces a slurry you can put on the heap for nutriments and water benefits all at once.

hints from municipal composting and sludge production

A few of the practices used in municipal composting give some good hints to home gardeners who want to do high-speed composting themselves. For example, it's been found that it pays to start with a seed compost of old materials of 5 to 10 percent. The moisture in the heap should be around 50 to 58 percent, and 50 percent oxygen should be introduced by frequent stirring.

When the moisture goes up to 70 percent it is bad because the pore space for entry of oxygen decreases. Then the pile cools, gets soggy and putrefying, with anaerobic bacteria taking over. The best pH is in the range of 5.5 to 8 to prevent undue loss of nitrogen. The heap will be acid at first (pH 5.5), and when it goes up to pH 8, you can consider that the processing is just about complete.

Fresh municipal garbage fed into big layered bins can be completely converted within 70 hours when all conditions are right.

sludge for your garden

Many towns now have sewage treatment plants where the end product is activated sludge, which is germless, dry, and smells like the floor of the forest. It is a good source of nitrogen and minerals. A few years ago, it was reported that 50 percent of the sludge in the state of Connecticut was being returned to the soil. This is half as good as the Hunzas and the Chinese recyclers, who omit the treatment plant and put their garbage, manures, and night soil directly on the land. Using human excreta is somewhat hazardous because of intestinal bacteria and parasites unless there is enough heat in the compost to cook them to death. The heat needed for killing off typhoid, paratyphoid, and dysentery is 150 degrees F. for a week or so. Without that heat, these coliform bacteria could survive all winter. In fact, ice preserves them. Salmonella bacteria can survive a month, unless cooked, and the hepatitis virus much longer. In this country sewage is given a 200-degree heat treatment before it is sold for fertilizers such as Milorganite. Any use of night soil buried in a compost heap is therefore extremely precarious, for the borderline of safety is too thin for an amateur to be able to control.

By the time the organic gardener gets heated and bubbled activated sludge, the nitrogen content is about up to what it would be in cottonseed meal—that is, 5 to 6 percent—and its phosphorus up to 3 to 6 percent.

Sludge has other nutrients, such as calcium oxide, magnesium oxide, sulfur, iron oxide, and ash. Its pH runs somewhere between 4.5 and 7.2, but higher when certain industrial wastes are in it. Trace elements come from copper sulfate, zinc sulfate, borax, manganese sulfate, or sodium molybdate. In addition there are celluloses, lignins, and crude proteins, and it is 25 to 36 percent carbon.

With this knowledge, almost any organic gardener will be willing to use sludge, especially when it is heat-dried. Another advantage is that it is slow to be nitrified. It is good for a new lawn, for instance, at the rate of one part sludge to two parts soil, mixed down to a depth of six inches. For an established lawn, put it on in the winter, one-half inch deep, and let it freeze and thaw. For gardens and farmland, season it for six months and then put it on before rain, preferably in the autumn. Sludge will not only provide the major nutrients, it will also improve the physical properties of your soil and increase productivity of some if not all plants. Spinach is not much benefited by it; oats and beans are somewhat helped; and beets and turnips are very much helped from heavy applications.

sludge . . . Sludge is recommended by the Connecticut experiment station for lawns, melons, squash, and other vine crops. (But a warning is given to add lime after a while if sludge is continually used.) Connecticut sludges have been heat-treated, thus destroying any dangerous bacteria, and this is normal treatment for all sludge like Milorganite that is offered on the market. Many untreated sludges have been pronounced all right for all uses except vegetables. If in doubt, cure the sludge you get by letting it sit on the ground in the open for at least three months. Also if in doubt, use it only for vegetables to be cooked, or for those whose edible parts are totally above ground. If still in doubt, apply composted sludge only in the fall and let it cure again on the soil over the winter. I'd recommend this for beets and turnips.

To compost sludge, put it with materials like grass clippings, sawdust, wood chips, leaves, weeds, and vegetable residues. Unfortunately, sludge may contain DDT and other undesirable runoffs or discards as well, which are not biodegradable; then even the usual purification in the compost heap won't help. Also, if you handle it with bare hands (which you are advised not to do), wash afterward with warm water and laundry soap. Ditto for blood meal, by the way, because of possible infectious fungi not killed by the heat treatment.

composting in Africa and the Orient

Composting in Eastern countries is often introduced for the sake of disposing of all domestic refuse—not only garbage but also animal excreta and street sweepings. Many towns and villages now adapt the Indore method of Sir Albert Howard, with results as successful as Sir Albert's, if not more so, especially where the weather is good and hot. The refuse used to be burned, or buried in pits, and the nutritive values were mostly lost to the community's farms and gardens. Sometimes the cow, horse, sheep, and goat manure was heaped up in smelly piles, which leached away, or the cow-pats, as in India, were used for stove fuel or plaster.

Now the people in many villages are taught a new practice of mixing their night soil, vegetable wastes, and animal manures to compost all together in their pits. In their hot climate, with three turnings, they can achieve compost in about a month. The villagers are taught to keep the mass well moistened, use plenty of fresh green materials, insert ashes from domestic fires, and keep the pits well aired, lightly packed, and protected from becoming waterlogged. They then see that the mass can heat up on the second day, and shrink and be free of smell by the fourth. If made so this happens, the pile should attract no flies or other vermin, and will harbor the helpful bacteria and fungi as needed.

night soil . . .

The costs, of course, are much less than they would be for municipal incinerators or sewage treatment plants. The adjustment from dumping is not hard for citizens to make, especially when they find that the results are so very beneficial. In five years, according to one report, compost production in a Nigerian village totaled 43,800 tons. That's a lot of fertilizer.

Where the night soil is too wet the villagers dry it by mixing it with dust, wood chips, sand, and sprinklings of dry dirt and lime before putting it with the other materials.

In Hungary everyone has to take his refuse to the city composting center. Night soil, garbage, refuse, the lot. Hungarian rules for composting are rigid and suggestive. Night soil is mixed with peat, composted for at least 21 days, covered with a layer of previously aged materials, and then mixed with other wastes and further composted, or sold. The final mixture has to come out, according to rule, with specific carbon/nitrogen and other ratios, and proper density and ash content. The Hungarian compost centers, as in all big municipal composting processes, remove all the iron, aluminum, and glass.

Such activities around the world show that composting is really believed in by many peoples, and is studied in great scientific detail. The organic gardener gets ideas from them for what to do for himself, and what to use as arguments for nonbelievers—especially about the rapid conversion and purification possible, and the superior nutritive values of this potentially plentiful fertilizer.

chapter five

the soil in your backyard lives

Everything I've learned about what happens in the soil I've found exciting. As soon as I began to see that soil is not just that flat brown stuff good to put houses on or bury things in, the whole earth all around suddenly became alive and full of wondrous events. I suppose our wrong notion of the soil comes from looking down at compacted, barren paths outside schools and across vacant lots or from seeing the bared profile of a gravel pit as we pass by. We see the lowest layer of rock, followed by layers of different-sized pebbles, gravels, and sands in the subsoil, and the finer silt and loam in the topsoil. The soil is not just that compound of minerals and nitrogen, phosphorus, and potassium we were once taught it was. Far from it. It is teeming with many forms of life, millions of underground organisms that are active all the time making physical and chemical changes in the soil, and in all the plants and animals springing from it. Scientists can now tell us about various kinds of bacteria, ray fungi, true fungi, molds and mushrooms, green algae, blue-green algae, and other plants; about yeasts, many different kinds of protozoa, earthworms, nematodes, centipedes, ants, other insects and their larvae as well as the moles, shrews, and all the small animals that live in the soil. They can also describe many (though not yet all) interconnections between the organic and inorganic events going on in the soil all the time and can tell us that there is streptomycin in cow manure and penicillin in mold, and that vitamins do come from bacteria. They now use the term *biochemical* (or more ponderously *colloidobiological*) for what's happening—even for the weathering of the bedrock into small particles (now known to be caused in part by acids generated by bacteria).

I like these terms, for they do away with false separations of the so-called living and nonliving, and they *soil . . .*

show a good advance from the NPK mentality which seems to view the soil as essentially dead. Plants and animals cannot eat lifeless minerals. But a miracle of plant life is its ability—not only above ground in the green leaf, but also below ground in the mystery of the dark—to transmute soil materials, both organic and inorganic, into living sources of energy and food.

learning about soil

Before going on to say what I liked learning about the soil and its millions of beneficial bacteria and fungi and the chemical reactions they promote in the cycles of soil life, I am going to stall a minute for a few words about the two ways there are of learning, and the two kinds of enjoyment you get out of it.

The first is learning how to do something that will work. It is somewhat superficial, but very satisfying to the

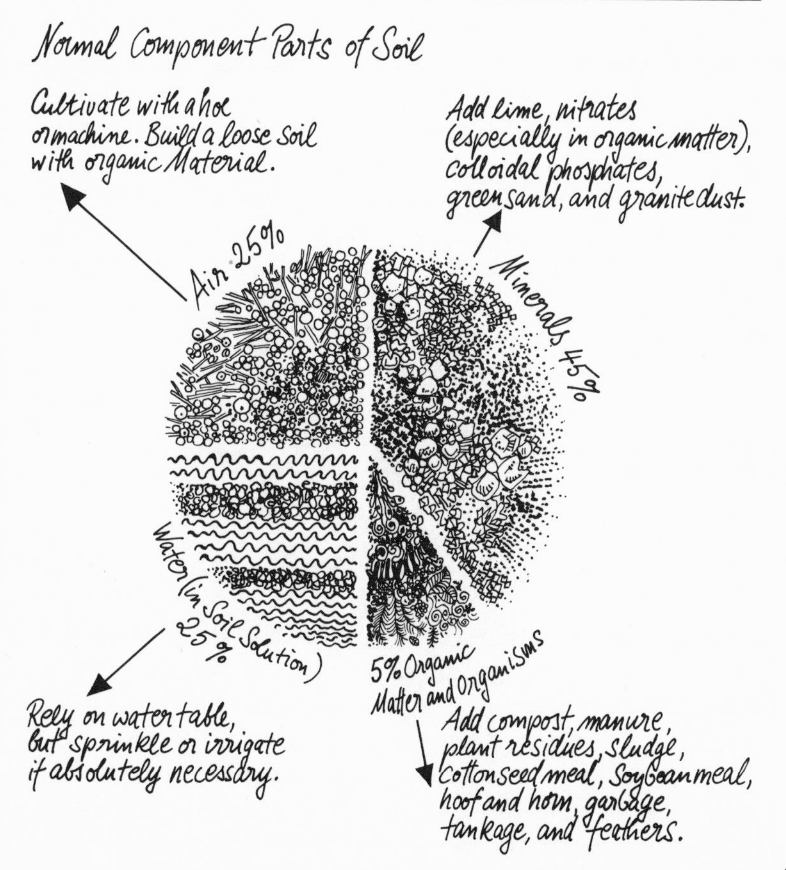

gardener when it brings results. The second is more probing, takes a lot longer, and brings understanding and feelings of clarity and wonder that are pleasurable and reassuring. You learn why it works.

The first kind is ancient, persistent, perennial, and quite satisfactory for many gardening situations. For instance, when it is a question of how deep certain bulbs should be planted, you can consult a chart with diagrammatic sketches of the bulbs of crocus, narcissus, tulip, and so on with marks showing whether they should be planted at 3, 5, 7, or 9 inches. The novice gardener goes ahead and does what he is told, and normally all will turn out well. You feel no need to know why, and you certainly aren't going to get embroiled in an argument with any chemical-political lobby working on Congress or over the TV to persuade you not to plant

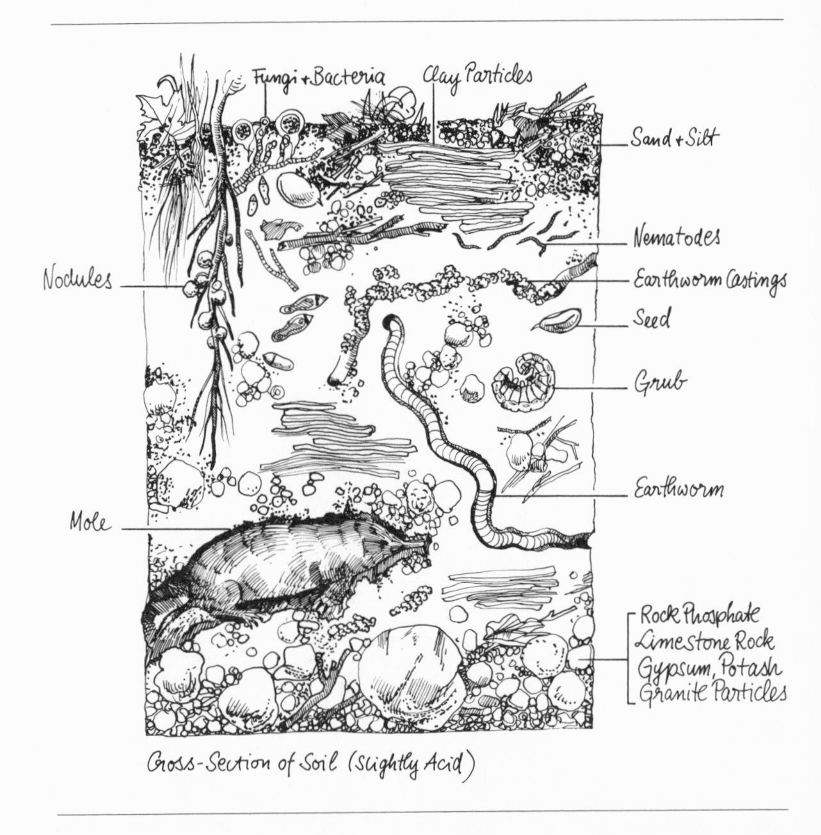

Cross-Section of Soil (Slightly Acid)

at such depths and to use other depths instead. You simply put your trust in the chart and have no temptation to do otherwise. You rarely stop to ask yourself or anyone else why.

learning about soil . . .

When it comes to questions of preparing the soil and of controlling pests that come to attack these bulbs once they are in the ground, however, you seek the kind of learning that gives you explanations. It may even change your attitudes. Bulbs are inviting to certain little pests. Some people rely on the first kind of learning and do what an advertiser tells them to do: use poison. Organic gardeners have attitudes that make us resist that kind of advice to use fast-working, hard chemicals because we want our own foods clean and a clean world for all living things from birds to bacteria (and humans who might be affected by the poisons). We are glad to learn that wire cages fastened around the bulbs before planting will deter mice, for mice turn out to be the pests.

As a novice you may not yet have had experience with a wire cage or its good results, but you can see the reason for it because a mechanical barrier of fine enough mesh will of course repel any little animal. You have already heard about the need for air in the soil, seen people scratch

Wire Cage around Bulb

it open to let air in, and you can understand that alleyways made by mice, shrews, and moles as well as earthworms are good things to have—so why kill these animals off with poisons? And why chance killing any of the beneficial microorganisms?

More complicated are the reasons you learn for having organic matter in the soil. When you hear it said you should put a lot of it into your garden soil, and accept the statement that it is good for the soil and the plants, a few first reasons are obvious—that bulky, airy stems and straw make the soil lighter, make drainage better, and make room for roots and worms to get through. A hard-packed, clayey soil clearly does not make room. You can see that spongy leaves, manure, and buried peanut shells will take up and hold water a lot better than a hard soil where gullies form instead. And you probably respond to the ethical appeal and the large reason that man should return to the soil whatever he has taken from it so it can cycle again and not be lost to the sea.

There are many other good reasons for having organic material in the soil, and the explanations are rather long, complicated, and biochemical. If all that follows is more than you ever wanted to know about what goes on, skip the rest of this chapter and go to the next chapter, which will invite you to do what you are told about planting, growing, harvesting, and cooking vegetables, herbs, weeds, and berries. For those who want to look into some of the whys and wherefores, read on.

the beneficial microorganisms

Man has for more than a hundred years known that armies of beneficial bacteria and other microorganisms are essential contributors to the life cycles that keep plants alive and thriving. Soil must support them and take part in the symbiotic relationships among plants and bacteria, fungi, larvae, molds and their antibiotics, worms and robins, moles, wasps, and many other creatures. Soil harbors hydrogen, hydroxyl, calcium, and many other ions, nitrates, phosphates, and many other chemicals as well as proteins, enzymes, and numerous other organic materials. As a gardener you influence and try to enhance all these interrelations and the conditions of air, water, and warmth that favor both microorganisms and plant growth.

beneficial microorganisms . . .

If your soil feels smooth, soft, and slightly oily, and if its pH is not too acid or too alkaline, you know you already have a good start. Unfortunately soil tests are governed by a chemical attitude to the soil and concentrate on the mineral content. They don't tell you what bacteria you need or whether you need fungi for the healthy plant-fungus relation (called mycorrhizal association) that provides a remarkable exchange of nutrients—the plant gives carbohydrates to the fungus and the fungus gives proteins to the plant. The fungus grows up between the cells of the root to bring this about. It does not kill the plant or parasitize it either.

Most good soil test reports do tell you whether you need to add lime or organic matter to lighten your soil structure. Where there is plenty of manure and vegetable matter subject to decay, there will be organisms multiplying to take care of it. There will be both the bacteria that don't need oxygen and those that do, including the indispensable nitrogen-fixing ones whose activities provide much of the nitrogen that plants use. This kind of beneficial bacteria you can buy if you want to, though I have never heard of anyone's soil-test report that advised it. They are called activators, and are sold sometimes by seedsmen along with the legumes which have nodules in which these bacteria work. (See p. 232.) In a good soil there might be billions of bacteria in a pound of dirt.

These billions of bacteria and other soil organisms feed only on dead materials. Under the ground they prey mostly on dead plant materials, but also on dead animal organisms, most of them microscopic. Occasionally other species of bacteria, fungi, and little nematode worms attack the live higher plants, but they are minor compared to the many others who act on soil materials, converting them to plant nutrients. Organic gardeners tell you you also need earthworms and the protective ladybugs and praying mantises. Anyone will tell you you need bees and birds to keep the cycles going that provide for the needs of the soil and the plants.

Scientists now know that the bacteria need roots, for they also use the organic foods that ooze out of the roots during their processes of metabolism. These interdependencies between higher and lower plants, among trees, flowers, and vegetables and bacteria or fungi, are the main reasons why Sir Albert Howard and his followers have all been so reluctant to hazard putting nonorganic substances into the *beneficial microorganisms . . .*

soil without understanding their full effect. It is not a simple matter of putting in one shot to hit one target. If you touch off one reaction in the soil, it may lead to a hundred (or eventually even a hundred thousand) others.

the soil air and soil water

Soil scientists have not yet shown conclusively what nonorganic substances do to the soil air and soil water. But they do know that the oxygen content of the soil atmosphere is essential for the metabolism of almost all soil life, for the exchange of carbon dioxide and nitrogen in the respiration of the roots of plants, and for nearly all the chemical reactions that go on in the soil water that films all soil particles and flows through the soil, touching all the minerals, all the organisms, and carrying everything into solution that will dissolve. Soil air also brings in some of the atmospheric nitrogen that is essential for bacterial conversion to ammonia, then nitrites, then to the nitrates which are available to plants. It circulates the sulfur dioxide, carbon dioxide, and some of the hydrogen that take part in other reactions, and opens pathways for roots to find their way to nutrients, and to aerate the soil solution. Rain enters, and when the soil solution dries up or leaches out, the air comes in again to fill the spaces. The soil solution is of course essential to transport nutrients. It is usually slightly acid compared to plain water, and this helps make it capable of aiding many chemical exchanges, including eating away rock minerals in the soil.

The minerals are the silicates, phosphates, potassium and calcium compounds, magnesium salts, and dozens of others. To the chemically minded these soil components are of prime importance. To organic gardeners they come second because no chemical reactions occur at all without the aid sooner or later of living biological agents.

the Liebig analysis

Actually such distinctions are merely academic, for the events of the soil are cyclic, or rather in a network, and all parts are needed for a plant to thrive or grow at all. When the chemical events in the soil were first discovered in the early nineteenth century, people jumped to the conclusion that the mineral content and the chemical reactions were the whole story. They threw over the long-honored theory that it was humus which made plants grow and rushed to adopt the new notions. When German chemist Baron Justus von Liebig came forth with the theory that soil is chemicals, this had a potent effect on agricultural circles, especially in England, where converts to the theory formed hundreds of farmers' clubs to study soil chemistry.

the Liebig Analysis . . .

What appealed to everyone was the clarity and simplicity of Liebig's notion that what plants needed was simply nitrogen, phosphorus, and potassium—N, P, and K. He demonstrated this by burning a plant, analyzing the ash, and stating his findings as mainly phosphorus and potassium. He said the third component, the nitrogen, had gone off into the air. Large deposits of phosphates and muriate of potash were opened up in Germany at about the same time. If farmers and gardeners would buy these salts and nitrates, all the needs of their plants would be filled, Liebig claimed. These, not manure or humus, were what plants essentially needed, he said. When the chemical converts were made, scientists had not yet discovered bacteria or that there were bacteria in manure as essential as the nitrogen in it.

Soon a demonstration station was founded at Rothamsted in England, and it is still conducting experiments to support these claims of the chemical nature of the soil. From that day to this many Englishmen, and many Americans, have believed that plant life, which men for centuries had thought of as a mysterious complex process, was really a neat, simple, chemical process easy to understand and control by applications of NPK.

But old-time, conventional horticulturists and estate gardeners as well as various plant scientists began to object. Out of their objections has arisen the organic gardening movement, especially out of the objections of Sir Alfred Howard at the time of his very practical findings about soil in India. Sir Albert and the leaders of the organic gardening movements in England and this country have attacked the logic of Liebig's thinking that since N, P, and K were the main constituents of the plant, all it would need as food to make it grow would be nitrogen, phosphorus, and potash. "Now, if the human body is burned up, its ash will be found to be similarly rich in these chemicals," wrote the late J. I. Rodale, founder of the organic gardening press, "but does that mean that we should take it in chemical form? Of course not. We get our N, P, and K in ham and eggs and in vegetable *the Liebig Analysis . . .*

soup. In the same manner, the soil should secure its NPK from living, organic foods, not from dead chemicals." Liebig's way of approaching plant physiology has seemed to many of us to be dead, lifeless, and wrong. We feel that analyzing after the plant is killed and burned is neglectful of practically all the living factors. Macabre as it may seem to us, many forces in England's rapidly industrializing society in the 1840's made the public of that day all too eager to adopt a view of a plant's requirements in terms of inert and simplistic chemistry. It made a plant seem like one of their new machines, and it spurred man on to invent more and more to plow, drill, harvest, and spread the new chemical fertilizers to manage the plant machine.

It also changed agriculture and horticulture from an art to a commercial enterprise and shifted the emphasis from quality to quantity. Once the machinery was there, it had to be used, and the spiral began. Machinery led to monoculture; monoculture led to feasts for pests; pests led to pesticides. Pesticides soon led to resistant pests, and then the hybridizing of plants that have to have pesticides. Pests learn to go for new hybrids, and trouble results again. Rodale commented, more than twenty years ago, "It is exceedingly dangerous to operate with only 99.99 percent of a formula." Today the effects of tiny amounts of DDT or mercury on living tissues are known by nearly everybody, and people are aware now, as never before, of the importance of one part per million, or even one one-hundredth of a part per million when it comes to the cause of illness or poison, or to the cause of health and well-being of living things. The tiniest trace of cobalt deficiency in the soil can cause plant trouble; minute soil deficiencies of iodine in a community can cause goiter in humans.

calcium

All such trace elements in the soil are essential, as are the more plentiful elements like calcium, which leaches out and has to be replaced by liming when man disturbs the soil.

Calcitic and dolomitic limestone are used; they both derive from sea salts or the ancient shells and bones of prehistoric creatures deposited on ocean bottoms before the continents were formed. The calcium in your bones, in your garden, and in the lettuce you ate yesterday has been cycling in and out of the waters, soil, plants, and animals in nature's ecosystem since those presedimentary eras. To bring back a leached-out calcium content, organic gardeners prefer dolomitic limestone, because it dissolves slowly and also has a valuable magnesium salt (needed for chlorophyll) as well as the calcium salts. Your plants need calcium for the cellulose in their cell walls, and the soil needs it for texture and for the ion exchange that is the basis of all biochemical events in the soil.

calcium . . .

Test your soil for calcium, for if your composting and fertilizing programs involve high-nitrogen materials such as lots of green grass clippings, pea or bean plants, or leaves, you may need to add lime to help slow down excessive releases of nitric acid. This acid would inhibit the nitrogen-fixing bacteria, and the garden would suffer. (The soil gets acid when the calcium ions are replaced by hydrogen ions, or rather when too many are. The calcium ions leach out, leaving the acid hydrogen ions to dominate. In humid, temperate climates like the Northeast, this happens a great deal, so we must be wary and put on lime when needed. There can also be an increase of toxic aluminum ions.)

You are lucky if you have limestone deposits on your land, or a cave with stalagmites. Both are fine sources of calcium.

phosphorus

The soil must have phosphorus, which is also a building block in all organisms. Unfortunately, many soils are low in this nutrient, and we consume and flush away tons of it that were in grain, milk and livestock and meat. Rock phosphates in pulverized form are the fertilizers organic gardeners use to return phosphorus to the soil. They prefer it in this form to superphosphate, which has had sulfuric acid added to it. Superphosphate makes the phosphorus rapidly available for plant use, but it leaves a residue of calcium sulfate.

In addition to rock phosphates you can get colloidal phosphates, but if you decide on these you have to use half again as much to get the same amount of phosphoric acid into your soil. (Mix it with manure, and apply 10 to 15 pounds per 100 square feet in both spring and fall.) Before the present sources of phosphates were found in Florida, Tennessee, Nevada, and the German mines, bones were used to satisfy the demand for calcium and phosphate fertilizers. Old Liebig complained in his day that England with her craving for fertilizers was robbing the graves of battlefields, in Sicily, the Crimea, even Waterloo, to use up, flush out, and "squander down her sewers to the sea."

potassium

A good potassium or potash (K_2O) supply is necessary for fine strong plant growth, for the transformation and transport of materials that are made into sugars, and to aid maturing and seed production. Potash deficiency means weak cell walls, weak stems, and slow photosynthesis. (One example you might see in your garden is the cucumber that stays narrow at the stem end, but bulgy at the flower end. If you see this, add a good mixed compost. Also spread some extra wood ashes, and mulch with buckwheat straw, dried cow manure, or alfalfa hay. Or greensand and granite dust.) Greensand, an iron-potassium silicate from under the sea, contains many nutrients, including 6 to 7 percent potassium. The related green glauconite, like kitty litter, has clay particles that absorb and store quantities of water. Granite dust, from quarries and stonecutters (or in a product from Georgia called Hybrotite), can provide from 3 percent to a whopping 11 percent of potassium. These dusts also contain feldspars and mica. They are cheap, long-lasting, and without any bothersome residues. Apply at the rate of 20 pounds per 100 square feet, preferably in the spring. Though potassium is already plentiful in the soil, plants can get only 1 percent of it.

summary of mineral supplements

The four sources of natural mineral fertilizers that appeal to organic gardeners, then, are rock phosphates, colloidal phosphates, greensand, and granite dust. Add them as I have said to manure and compost, keep air in your soil, give it good drainage, and protect it from drought. Then you are well on your way to excellent loam soil—if it has enough nitrogen.

nitrogen

Nitrogen, the third major nutrient (which is usually listed first), is derived almost exclusively from organic matter, dead or alive. You have a good start or a bad start with nitrogen depending on where you live: if you are in a prairie state, the soil is still rich and high in organic content with an average of four tons per acre down to a depth of three feet or more. If you live in the Southeast, however, you may have only one ton per acre, and in the Northeast little more than that where the leaching and erosion have been considerable. In the air over every acre of land there are tons of unusable nitrogen, maybe 80 percent of the air.

nitrogen . . .

Almost every cycle in the soil-nutrient process depends on the nitrogen cycle, which depends on two kinds of bacteria that change the nitrogen compounds into forms that plants can take up and use.

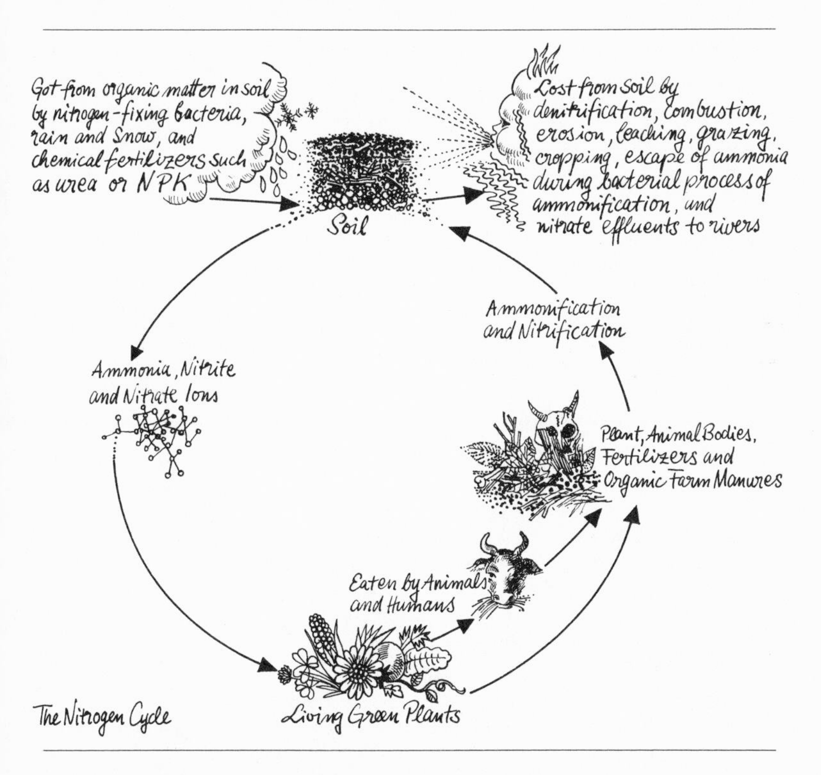

The Nitrogen Cycle

The bacteria living in the nodules on the roots of legumes such as clover, peas, or beans are capable of combining nitrogen with carbohydrates they get from the legume plant to form proteins, with the energy they use coming from the carbohydrate. This produces enough nitrogen compounds for both bacterial and plant use, probably as amino acids. It happens only when two such organisms as higher plants and bacteria are involved.

Aside from these symbiotic bacteria, there are free-living ones (both aerobic and anaerobic) which transform up to 40 pounds per acre of the carbohydrates from organic materials to proteins. And some bacteria break down proteins, form ammonia, and then nitrites. Others then form nitrates, and all are used by plants. When they manufacture food from raw materials, these bacteria are mysterious and *nitrogen . . .* potent as the cells in the green leaf which make food there. Those cells use sun energy; the bacteria use chemical energy which they get from the compounds they break down.

Aside from domestic manures and other organic materials such as cottonseed and blood meal, commercial nitrate fertilizers needed in the soil have traditionally been supplied by guano or bird manure from South America and various other places. Today, the large source of nitrogen for commercial fertilizers is from the air, in a product called urea. In pure form, this is acceptable to organic gardeners, I have heard, but they object to the kind called ureaform, because of the chloroform said to be added.

carbon

Carbon is so ever-present and so plentiful that it is rarely discussed much by gardeners building up their soil. It makes up about 50 percent of all organic matter, and is the base of many organic constituents, including carbohydrates.

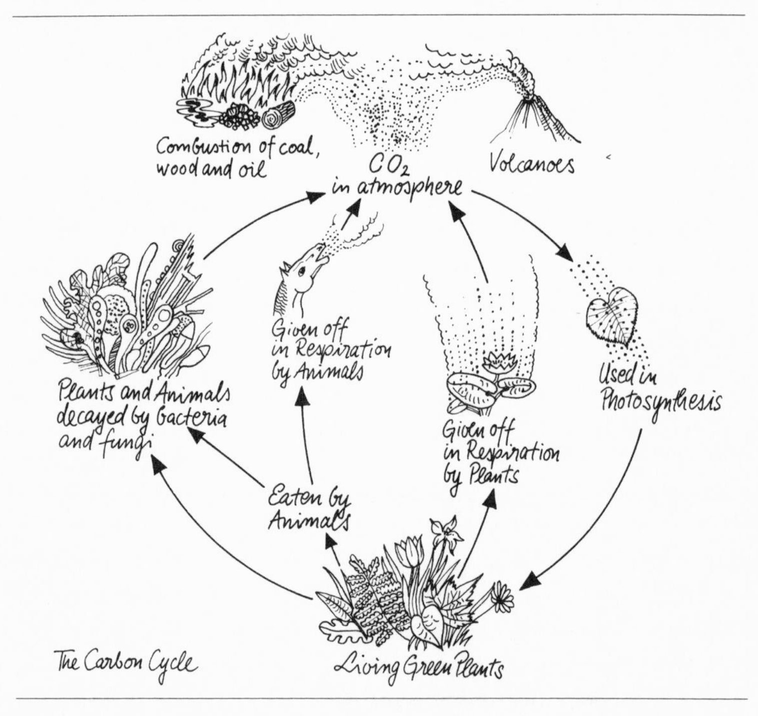

The Carbon Cycle

Though pure in a diamond, it practically nowhere else is. In the oceans, in the air, and in the soil atmosphere it is plentiful as CO_2, carbon dioxide, and it is present in such familiar substances as washing soda ($NaCO_3$, sodium carbonate) and limestone ($CaCO_3$, calcium carbonate). Every cell of every plant and animal uses carbon, but it constantly recycles through organic processes, breathing, death, and decay. Decay organisms give off most of the carbon dioxide that replenishes the world supply.

carbon . . .

Sulfur and magnesium are the other two more or less major elements in the soil. They are needed for plant structure, both as an element in the plant cell (magnesium) and as a process aid (sulfur). Magnesium, in fact, is the central atom of each chlorophyll molecule.

trace elements

The trace elements in salts of manganese, zinc, copper, iron, boron, and molybdenum and sometimes cobalt, iodine, and fluorine, as needed by plants, are found in nature in manures, humus, and all sorts of organic wastes. They are one good reason these materials are used by organic gardeners for recycling soil components from one organic form to another.

In the lab or on your windowsill you can notice deficiencies of trace elements in your plants, but outdoors in the average composted garden you are not so likely to. Celery deficient in boron will have brownish spots, beets will get mushy or corky, cauliflower will be stunted and hollow-stemmed. The tip of the growing stem of tomatoes might get black. Iron deficiency also causes spots and yellow leaves. Copper deficiency will make lettuce look bleached and onions very thin-skinned. A shortage of zinc will cause yellow, mottled leaves, noticeable particularly in mustard, tomatoes, and squash. If it is manganese that is deficient, the plants will grow slowly, mature unevenly, and become yellow between the veins of the leaves. The soil you provide, therefore, should be rich in all the elements your plants will need to grow green and strong. It will be if your composting is halfway adequate.

why we treat the soil

The organic gardener believes today more than ever before that what one does to the soil should be done as part of a total process of cycling. We do it knowing that protection and reclamation have become essential. Now we are alerted and alarmed about the damage we have done to our land and waters and wildlife by careless exploitation. We know that bison and passenger pigeons do not reappear if wiped out. The soil, once depleted and eroded away, will take centuries to regain its topsoil without armies of composters and tons of organic matter reclaimed to help it come back.

I think all this enters into the desire of organic gardeners and the new conservationists to do something more *why we treat the soil . . .*

sensible and more moral with the land. A busy and conscientious organic gardener can aid in the reclamation of topsoil and, aside from the benefits in better food for the family, feel also this moral obligation to enter the process and contribute to stemming the downward path of planetary exploitation, degradation, and depletion.

We are also aware that terrible side effects of depleting the soil are the impoverished and ill-nourished farmer and his family who have had to remain on worn-out land. The rundown conditions resulting from thoughtless, ignorant, or avaricious practices rarely if ever were followed in America by efforts to restore what had been lost, let alone to prevent it. People just moved on; and as you can see when you cross the country, there are many places where the subsoil has actually come to the surface. The rich virgin topsoil disappears, its nitrogen content gone down to about zero, the mineral nutrients nearly exhausted, and the microbial life quite vanished. We know that people trying to subsist on such soils are appallingly malnourished. Their plants achieve no proper size; and they are very deficient in proteins, vitamins, and other essential sources of nourishment.

what a plant gives to the soil

A good healthy plant will keep on making and providing organic nutrients for itself and for the soil, from its roots. It makes sugars, carbohydrates, and starches. By another route the simple sugar made in the leaf builds to oils, and when nitrogen is added it builds to amino acids and then proteins. Not sent out into the soil from the roots are the cellulose and woody lignin fibers. All of these and some of the other products such as vitamins, enzymes, gums, even alkaloids, are necessary for plant, animal, and soil life. The enzymes, vitamins, and antibiotics come and go, back and forth. In the patterns of interdependence, for instance, soil bacteria evidently supply vitamins as well as proteins to plants, and such molds as the penicilliums provide the antibiotics. All of these soil processes go on between the temperatures of 40 and 100 degrees, and the number is stupendous.

the new analysis

Today when scientists dry out plants and reduce them by a dehydrating process different from old Liebig's simple burning, they find that 95 percent of the dry weight of a plant is material that has been organically synthesized. Half of that is carbon. They also analyze sap as it comes up the stem from the soil solution and find that its mineral content is only .03 percent to .13 percent of the total. Many now believe that the actual role of these minerals is to be catalyzers in the organic processes. The sap also contains organic compounds that it takes up from the soil, about as many again as the minerals.

the new analysis . . .

Long before all these discoveries were made about the living elements in the soil, and long before antibiotics or enzymes were thought of, the battle was raging rather bitterly between the NPK advocates and the organic gardening advocates. One side says that chemical fertilizers—and the pesticide poisons they and other agricultural practices necessitate—are bad for the cycles of life in the soil. The other side says that soil exhaustion by cropping, leaching, erosion, and so on makes it obligatory to replenish the loss with quick-acting supplies so farmers can feed livestock and people. Unless paid not to use their soil to raise crops, farmers know that up to a point they can make bigger, faster profits from using nitrogen fertilizers (or the so-called complete NPK fertilizers). Published for these farmers in books like the United States Department of Agriculture yearbooks are charts indicating exactly the points of diminishing return—when it is no longer economically feasible to add another ounce of nitrogen on the corn crop, for instance. The contention you can hear is that a world food supply would never be possible without pesticides also. (See Chapter 7.)

The agonies that organic farmers and gardeners began to have in the early part of this century when they saw the disappearance of earthworms, for instance, from artificially fertilized plots are now agonies ten times worse since the introduction and use of millions of tons of those pesticide poisons. They deplore the spiral of chemically grown plants needing sprays, needing new varieties, and so on. Now the southern field corn blight in the Middle West has shown exactly what has been caused. We can only wish that these unfortunate farmers had used organic methods like composting, mulching, rotating, and organic sprays where needed. But corn is a fast-money crop, and even the wisdom of rotating has all too often been neglected. The farmers say their county agents, schooled only for corn in those areas, did suggest leaving stubble for organic matter, but this has been more of a bane than boon when the infestation came from the South and began to spread.

the new analysis . . .

This situation is tragic. The race is on to breed a new corn again. It will be found, and perhaps a new spiral will be begun. Perhaps, also, the period of distress will lead to further reconsideration of the methods of fertilizing and health protection that should be used by the farmers and gardeners who want good crops and a healthy environment for them.

one ray of hope

We are not very hopeful. But the experts at Cornell and at other outstanding agricultural schools in this country are now again stressing that organic matter in the soil is good for several reasons. They remind farmers that it granulates the soil particles, making a loose and friable structure; that it is a major source of nitrogen; that it increases the soil's capacity to hold water and helps to regulate the plant's water-holding capacity and the flow of the soil solution. They also say organic matter in the soil is the main source for energy for the soil organisms, and that without it biochemical action in the soil would stop. They point out that the humus particles are those that can hold both water and nutrient ions (and can do it better than the clay particles, which also promote ion-exchange). In saying this they underscore the fact that humus or organic matter in the soil contributes in essential ways to the soil chemistry. In addition, it produces, through the action of the bacteria who live in it and work it, that odd gelatinous substance which makes the lumping of soil aggregates possible. In a few years these professors will probably add such other values as the organic antibiotics and nutrient values. "Indeed it can even be said that practically all natural soil reactions are directly or indirectly biochemical in nature" is the cautious word of the experts. The present focus is on the reduction of residues to humus by soil microorganisms, and the consequent colloidal particles as major controllers of both chemical and physical properties of the soil. Their conclusion from all this now leads the authors of the leading agricultural text to depart from their ancient stance and to organize their study from the "colloidobiological viewpoint."

It is heartening that the agricultural schools are beginning to take this stand, and that the textbook used at many universities now expresses it.

chapter six

vegetables, herbs, and wild plants

So far my message has been: Make compost, understand it, and use it and other organic materials for the sake of a good soil, good nutrient cycles, and a well-protected garden.

This chapter will now get down to details about your vegetable garden. I will also recommend in my alphabetical list some useful flowers, weeds, beneficial insects, herbs, and recipes, because I think all those things go together to make you and your garden organically interdependent. Here and there I'll recommend biological controls, dusts, sprays, birds, and pest traps to use so your life as a gardener will be happier.

I do not want to entice you into any belief that gardening is foolproof. With luck and good mulch, it can be fairly easy, as Ruth Stout is fond of preaching; but a gardener can never get the idea that he has mastered the tricks, or that he absolutely knows what he is doing. With ingenuity and care, you can grow a fine garden and probably achieve the second best; but who ever grew the absolute best? I am reminded of the sentence used by the humble and pious Dr. Ambroise Paré, the surgeon who traveled with the sixteenth-century French army caring for the wounded. After each episode he narrated, he added: "I dressed him and God cured him." A gardener can but hope for equal success in the dressing of his garden.

You know you want to grow good vegetables and that the freshest, young vegetables are not only the most delicious you can have, but also the most healthful, especially if picked just before sundown at the end of a sunny day.

Studies of vitamin C and other nutrients in leaf vegetables and in mulberry leaves picked for silkworms have now revealed that picking leaves in the late afternoon or at dusk is indeed the scientifically correct time for harvest. Nutrition values are highest then. Vitamin C can be 20 percent higher in the leaves of plants in the evening than it is *aging of vegetables . . .*

early in the day. Proteins and sugars are also in larger supply after a whole day of sun, and especially at the end of a sunny spell of weather.

The younger the plant, when harvested for food, the higher is the protein proportion as compared to the carbohydrate. And the lower is the cellulose roughage. You have to have your own plants to get young ones. And besides, a vigorous young plant, not yet at the age of being senescent, is a disease-resistant, pest-resistant plant. It does not need to be sprayed. Only the weaklings get pests.

Actually, aging of plants in the sense of letting them get to the stage of becoming pest-susceptible is not just a matter of a number of days. Vegetable plants allowed to stay unthinned and crowded will get old and sick in a very short time if there is a rainy or moist cloudy spell. When it's wet the fungi get in, and the earwigs and slugs, and you can have a fine mess on your hands in a few days.

It will be a gooey mess at first, and then when the rain stops and the sun comes out, you will have brown, dead, no-good lettuce, for example, wherever there was crowding and decay. Another cause for susceptibility can be excessive dryness, then starving.

Along with this kind of aging goes stunting, and it, too, can come from neglect of thinning. Carrots or radishes, for example, which are not properly thinned will be skinny, stunted little plants with no root development to speak of, or with crooked and distorted roots. In either condition they, too, are subject to attack by pests that come to bite into any warped or abnormal spot they find. Once pests find a welcoming environment, they can multiply and then really cause trouble.

A clean garden, properly thinned and aired—if the soil is well composted and rich with plenty of organic matter in it—is really and truly more likely to be a healthy garden. Organic gardeners are repeatedly testifying to this. The microorganisms in rich soil prefer dead organic matter first in their diet when they can get it, and so they clean out those materials and leave your live plants alone. Slugs and snails also prefer to live on dead plant matter, attacking only the living parts of plants after dead materials have become scarce. (Pick off dead leaves and toss aside for traps.) Earthworms eat as much organic matter in the soil as they can get, and there are many more scavengers who stay in their place as long as there is the food they require.

preparing vegetables . . .

This is why we use biological aids and controls so as not to harm such useful creatures.

You can grow nearly the best vegetables in the world and then ruin them by the way you cook them, or the way you maltreat them before cooking. If you want to preserve the high quality you have attained by your good growing methods in the garden, study what happens to the original food values in vegetables from heat, light, pressure, seasonings, and milk or water.

We have all heard that freshly picked vegetables, rushed to the boiling water and cooked a very short time, are the best vegetables. We judge this to be so because we know they taste better that way.

Now scientists have studied all the changes and have assured us that the vitamin content, as well as the proteins, sugars, and minerals, are all protected by methods that also favor a better-tasting vegetable. For instance, if vegetables, after picking, are left in the light and warmth, they lose half their vitamin C content and much of their folic acid and vitamin B in a few hours. It is only sensible, therefore, to go to the garden with a paper bag or other dark container and pop the vegetables right in it. And only sensible to put the bag in the refrigerator as soon as you get back to the kitchen. Never soak. If you do, you may lose 75 to 100 percent of the sugars, minerals, and water-soluble vitamins. If you have chopped or peeled the vegetable, you will stand a chance to lose all three, whether you cook the vegetables or not. So go easy on peeling and chopping. The aromatic oils are lost too, as you remember from the taste of the gray, overcooked peeled vegetables that you have eaten in the past. Even letting vegetables sit for a few minutes drenched in the water you washed them in will leach out good flavors and good nutrients. Always wipe or whirl in the air immediately after washing. If you don't whirl, the water that clings to them will dissolve out the sugars, iron, much of the vitamin C, and other nutrients. Then chill, unless you are putting them right into the oil in the pot or the oil of your salad dressing.

Do not salt. Salt draws out the moisture and more of the dissolved vitamins and minerals. Salted spinach will lose nearly 50 percent of its iron, something which Popeye never told us. Braise vegetables in as little water as possible, and save every bit of cooking juice from vegetables to use in sauces, soups, and drinks.

cooking vegetables . . .

A good method (taken from old Japanese and Chinese practices) is to sauté the vegetable in a little soy oil, corn oil, or butter, simmer it slowly in the oil and its own juice, and only add enough water to keep it from burning. Since oxygen depletes the vitamin C content, before you put the vegetables into the oiled preheated pot you can add a few tablespoons of hot or cold water to replace the oxygen inside the pot by steam. Contact with oxygen will then be cut by both the oil and the steam, so loss of vitamin C is reduced. If you have hard water, add a drop or two of vinegar to reduce the alkali. Better still, cook your vegetables in milk, at about 200 degrees, for proteins combine with both acids and alkalis. Or use soybean juice, for that is protein, too. The vegetables will taste very mild and pleasant, and you will have more of an urge to keep the juice for soup. Keep a lid on the steaming vegetables while cooking, and keep the heat low.

All this shows how you can save in health, taste, and eventually cost, for this method of cooking vegetables takes very little heat. (Savers of electricity, take note.) A pressure cooker, which is what I use, will give the same results. Put the vegetables into ¼ cup or less of hot water. Peas take about a half a minute after 15 pounds of pressure is reached; asparagus takes a minute; beans perhaps two minutes; and even hard things like brown rice or soy grits or soaked wheat berries take only twenty or thirty minutes.

The top of a double boiler is probably even better. Have the water hot in the bottom part. Put a few tablespoons of oil and water or oil, milk, and water in the top, add vegetables, and bring them to a boil over direct heat. Let them cook a few minutes so that the enzymes that change the nutriments are destroyed, and then return the top to the base of the double boiler and simmer until done. Remember that after being heated through, vegetables do not need to go above 206 or 210 degrees to cook.

If you bake vegetables, presteam them a few minutes to stop enzymes acting and thus the loss of vitamin C. Leave on the peels and oil them if possible. When adding vegetables to stews, always have the stew juice hot first. Keep the lid on to prevent loss of steam and contact with oxygen.

If you do wish to cook shredded or cubed vegetables, chill them, then put them into hot oil, and sauté them quickly. Even people who do not like root vegetables find carrots, beets, and turnips very palatable if cooked this way. You can add milk or water, if you wish, or dip in batter first.

cooking vegetables . . .

For further flavoring, add a little more oil just before serving. Or add some butter. The oil is better for you than butter or margarine. Corn, soy, olive oil, or a bit of mayonnaise will do well. To cream vegetables, cook them in milk, and ten minutes before serving, add a thin paste of flour rubbed into milk in the proportion of 2 tablespoons to 1 cup. Do not let it boil. Vegetables recommended for slow cooking in milk are: Jerusalem artichokes, asparagus, beets, carrots, celery, celeriac, spinach, snow peas, shredded string beans, kohlrabi, mushrooms, peas, cubed summer squash, and all greens, for their acid or bitter flavor always disappears if cooked in a protein like milk. Or make a cream sauce, and then simmer the greens in that for about eight minutes. In addition to those you are familiar with, try watercress, parsley, radish tops, endive or escarole, celery leaves, cauliflower leaves when young, broccoli leaves, kale and cabbage outer leaves. Sorrel, Swiss chard, beets and beet tops, and turnip tops also do well in milk or cream sauce for gentling them. Or cook first in 1/4 cup of milk and then add sour cream.

If you cook your food any old way, and throw half the nutriments down the drain, what is the use of going to all that trouble to grow high-quality vegetables in the first place? The least you can do if you think these methods are too fussy is to eat most of your fruits and vegetables raw.

edible weeds

Grow whatever herbs, berries, and edible flowers your land will permit, and leave room to cultivate some of the edible weeds which can make a nourishing, savory addition to your menus. Such classes of plants are grouped together in the following list for the reader's convenience, though an occasional outstanding culinary treat like the *day lily* can be found in the main alphabetical list under its own letter. The *dandelion,* which is either wild or cultivated, can be found there in the main list too, as can *geranium* (as a pest repellent) and the rose that is richest in vitamin C, *Rosa rugosa*. The standard vegetables that will appeal to most gardeners as essentials for their larders are given fullest treatment; others only get a few lines for some special interest or entertainment. The cross references here and there will help you find the things you are looking for. So will the index.

Some Pests..........
Tent Caterpillar and Moth
Aphids
Leaf Hoppers
Wire Worm
Caterpillar
Cabbage Butterfly
Asparagus Beetle
Spider Mite
Mexican Bean Beetle
Japanese Beetle
Cucumber Beetle
Cabbage Worm
Stalk Borer
Cabbage Worm
Cutworm
Chinch Bug
Earwigs
Grubs
Sow Bug
Flea Beetle
Grasshopper
Millipede
Snail
Slug
Potato Beetles
Nematodes

... and their Natural Controls
Lacewing
Praying Mantid
Ducks
Honey Bee
Ichneumon Fly
Ground Beetle
Wasp
Ladybug
Garlic
Onion
Rove Beetle
Ants
WOOD
GARLIC SPRAY
BONE MEAL

the list

acerola
malpighia glabra

Be sure to grow it if you live in Puerto Rico, for it is a native cherry-like tree there. There is 85 times as much vitamin C in its juice as in an equal amount of orange juice. If you live in the continental United States, go to your organic food store and buy a candy bar made of acerola and carob. Deliciously tart.

achillea

See *Yarrow.*

ants

You will have ants in your garden, and you can value them for the hundreds of larvae of fruit flies and house flies that they eat, as well as caterpillars in orchards and some of the insect pests of the forests. Chinese orchardists like ants so much they provide little bamboo highways for them from tree to tree. If you don't want these creatures to walk into your kitchen, plant mint and tansy by the door to drive them away.

apples
malus pyrus

Buy two- or three-year-old scab-resistant varieties of such trees as Grimes Golden, York Imperial, Baldwin, or Jonathan, and do not plant on hardpan or gravelly subsoils. The topsoil should be very fertile, dressed with lime, compost, and fish emulsion and given plenty of earthworms. A legume cover crop, especially around the trees, helps. So does mulch. The location must be well drained, protected from harsh winds, and up from cold valley areas where the blossoms might get nipped in the spring. Also avoid any barriers that block the downhill flow of air.

For cross-pollination, you should plant two or more varieties of apple trees. For highest vitamin C choose Calville Blanc, Sturmer Pippin, Northern Spy, and Baldwin; the highest concentration of vitamin C is just under the skin, and in the skin itself—which has five times as much as the flesh. The dwarf will bear in three to five years as compared to eight to twelve for standard-sized trees. Plant in the fall after a summer's cover crop, and prepare the ground very thoroughly beforehand. Later, fertilize each tree once a year at the drip line (just under the ends of the branches all the way around the tree). Blood meal is recommended, at a rate of 2½ pounds per tree.

If you spray, use only dormant oil spray of miscible oil and water, early in the spring before the buds come out. Catch the apple maggot fly with traps made of jars filled with one part blackstrap molasses to nine parts water, plus a yeast cake. Let this ripen 48 hours. Hang from a branch. Will catch 100. One ounce of ryania in 2 gallons of water will sicken codling moths. Flush out borers with boiling water, or prick them with a wire. At all times practice stringent sanitation, and clear up absolutely all debris that might harbor disease or insects.

Test your soil once in a while, and if acid, apply a pound of lime and half a pound of rock phosphate per tree. Avoid simple sod around your trees. Use alfalfa, first as a cover crop, then as a mulch. Try to keep the trees trimmed and well ventilated. Cut out suckers.

Make apple pan dowdy at least once a season, cider and apple wine if possible, and make apple butter. To make your own apple pectin for all jellies and jams, wash a pound or more of unripe apples, cut them into thin slices, and add a pint of water per pound. Boil slowly for about 15 minutes and then drain through cheesecloth or nylons and keep the juice. Reheat the pulp with another pint of water and boil 15 minutes. Let stand for at least 10 minutes, then drain. Mix the two juices. Now you have pectin which can be used immediately with strawberries, pears, cherries, mulberries, blueberries, and elderberries—whatever you are preserving—or it can be cooled and saved for another day. It is especially useful for berries and fruits that do not jell well with their own pectin.

apricots

prunus armeniaca

The tree that kept the Hunzas healthy and happy for sixteen centuries. Grow if you can in your growing zone, or be bold and try wherever you live if you have a warm, protected place. Send for young trees. Write to the Organic Experimental Farm in Emmaus, Pennsylvania, to see whether their Hunza apricots did grow. Ordinary apricot trees are very tender. Plant in deep, rich soil in a northern but protected spot to delay budding until after danger of frost. Recommended: the Russian or Chinese varieties. Try dwarf ones. Be sure to plant two.

artemisia

artemisia absinthium

This is wormwood; use it for a deterrent. Plant it along with camomile and spurge to fend off mice and moles; mix a spray in the blender of one part artemisia to ten parts water for weevils, and add half a clove of garlic for a spray for other pests. Remember that it's poisonous.

artichoke, globe

cynara scolymus

Be faithful to it, in some fashion; it is a thistle-like but wonderful plant, needing lots and lots of room, usually on the West Coast. It should be tried on the Gulf Coast, too, and the Atlantic up to Massachusetts, but not inland. Perennial. Grows up to five feet given good soil well composted. In winter cover the crown with a bushel basket filled with leaves and manure. Start by buying two-year-old shoots or suckers. Clip off any dead leaves or roots before planting, and plant six inches deep. After two or three years of good heading, it may retreat. Propagate new shoots; or send away for more. Make rows 8 feet apart, and set plants 6 feet apart. Cut 1½ inches below lowest bracts, just before flower opens, and don't expect during the first year or so that any besides the terminal bud will amount to much. Cook promptly after cutting, though this is one vegetable, being pretty tough, that does not lose very much nutrient by standing. Clean a little under running water and pop into boiling water seasoned with juice of a lemon and cardamon seeds. Cook until a leaf pulls out easily (perhaps 25 minutes) and serve hot or cold with hollandaise, spiced mayonnaise, or French dressing. In early spring, for hors d'oeuvres, boil very small artichokes with a clove of garlic, a tablespoon of peppercorns and ¼ cup of vinegar. Chill, then remove outer leaves and the choke. Fill with chopped egg mixed with mayonnaise and parsley, lightly salted or chopped shrimp, or crab with chive.

artichoke, Jerusalem

helianthus tuberosus

A great favorite with organic gardeners, the Jerusalem artichoke does best in a cold climate but will grow almost anywhere. It is related to the common sunflower. One eats the roots, or rather the tubers, which have a nutty flavor. It is starchless, for it stores its carbohydrates as inulin and its sugar as levulose, not as fattening starches and sugars. Provides good vitamins—thiamin and pantothenic acid—and potassium. Plant 2 feet apart in April or May and cultivate as you would potatoes. Harvest any time in the fall and right through the winter. Mulch well for both easy digging and protection in winter. Keep moist in a crisper or cool cellar after harvesting until you eat the tubers. Replant what is left *artichoke, Jerusalem . . .*

over. You can grow 15 tons an acre, if you or your organic food store wants that many Jerusalem artichokes. (A good potato yield is 3 or 4 tons an acre.) They are free from disease and very prolific, will spread anywhere and could fill up your garden. To cook: Boil and serve with herb butter made by mincing 1 teaspoon sweet basil, 1 teaspoon parsley, 1/3 teaspoon thyme, 1 shallot, 1 small clove garlic, 1/2 teaspoon grated lemon peel, dash of cayenne pepper mixed with softened butter and rehardened. As usual, the best advice is to eat the vegetable raw and lose no vitamins. Scrub, peel off the dark spots if they bother you (or learn not to let them bother you), and serve with lemony mayonnaise.

asparagus

asparagus officinalis

There is nothing like the first sprouts of asparagus during the first days of spring, cooked within six minutes of picking, before the nutrients begin to depart. There is no sense cutting below the ground as commercial growers do. I feel the stalk, and break it just where the stem turns from hard to tender. If you do get some hard cuticle, you can slice it off just before cooking, but for maximum food benefit, do not peel and do not cut up. What tender tops cut like this you don't eat and freeze, send to your organic food store. Customers will be glad to get such a pure food product with no waste.

Steam it standing up in a little water, or give it a few seconds in the pressure cooker. Have water boiling to save vitamin C. Pour asparagus cooking liquid over toast (save vitamins, proteins), add hollandaise or melted butter. Very rich in vitamin A: six stalks give 500 International Units. Plus as much thiamin as an equal amount of rye bread or lean beef and half as much B_{12} as spinach, but twice as much as whole wheat bread or Swiss chard. Vitamin C is 15 milligrams in six stalks, compared to 25 in a tablespoon of lemon juice. You need 70 (female) or 75 (male) milligrams a day, so asparagus is a good start for the day, if you eat it for breakfast, as I do in the spring. Six stalks of asparagus also provide a goodly 160 milligrams of potassium. If you get totally bored with all other ways of serving asparagus, try the kind Alice B. Toklas once recommended. Tie asparagus in bundle, plunge in boiling water for a few minutes, then leave to steam for 6 to 8 minutes more. For a pound of asparagus, melt 4 tablespoons butter very slowly, add the cooked asparagus, still tied, and 4 tablespoons heavy cream. Turn so asparagus gets coated. Do not stir. Put on plate, cut string, add half a cup of whipped cream with 1/2 teaspoon salt mixed in. Whiz to the table before the whipped cream melts. How does that sound? You are lucky if you have an old bed that is well limed and well weeded (which you can harvest for four or five weeks). A new bed shouldn't be harvested until the plants are three years old.

For a new bed: Some of you may want to start with your own seed—Martha Washington for rust resistance. But I would advise two-year-old roots because growing from seed is difficult and takes three years. Set roots in trenches 12 inches deep and 12 inches wide, dug early in the spring, filled 4 inches with a very rich soil of well-rotted manure and compost dug into the bottom of the trench, and rained on at least once. Set roots in holes scooped out of this, about 18 inches apart. At first, sift 2 inches of soil over them and water them. As stalks grow, keep adding more rich, sifted loam and water, about every week. Do not plant near trees and big shrubs. Weed, and lime yearly. Also manure well. For extra protection, use a winter cover crop of soybeans or cowpeas. Or between the rows. Plant parsley, tomatoes, and rue over the bed for repellents. Be glad of moles and skunks that eat Japanese beetle grubs. Plant white geraniums if those pests get in. Even the USDA now recommends no chemical pesticides because they kill off the normal parasites on the asparagus beetle. Permit the parasite to grow. If you keep hens, turn some onto the plot. Or ducks, or guinea fowl, and keep the house sparrows coming with chick feed and bread crumbs. Pick off spotted cucumber beetle and dust plants with bone meal and rock phosphate (both good fertilizers, anyhow). If you see a diseased top (crooked, stunted), cut it off and burn it. It might have rust. To fend off the cucumber beetle, try planting zinnias, asters, and especially nasturtiums, all of which can be started indoors in early spring. Grapes and white roses can be used to attract Japanese beetles away from your asparagus—and make a garlic spray and chase them off the roses and grapes. Pick them off and drop them into a jar of water with a one-eighth-inch layer of kerosene or gasoline on it.

basil

See p. 159.

beans

phaseolus, vicia and dolichos genera

Bush beans, green beans, string beans, snap beans, or whatever you call them are such favorites, and are so easily grown, that some people plant six or seven times a summer—from late April to late July for a last fall crop. You need two pounds of bean seed for each 200 feet of row. Many beans will boost the nitrogen you have, for they are legumes, and harbor beneficial bacteria in nodules on their roots. Sow thinly, in rows 18 to 30 inches apart, with a hill of garlic every 4 or 5 feet, at alternate spots from the garlic in the row opposite. If that's too much garlic for you, vary it with onions, shallots, leeks, or scallions, but garlic has the most antibiotics. These onion-family plants not only repel insects, they ward off woodchucks and rabbits, too, especially if planted around the edge of the garden. For prettiness, and ease in picking, plant some wax beans and some purple ones. In food value they are no different. Remember to cook them whole as soon as you pick them to save nutrient values. Beans provide vitamins A, B, C, and G. We serve the purple ones raw as hors d'oeuvres, and get no complaints. As with other vegetables, there are more nutrients if eaten raw and if picked at the end of a sunny day. It is imperative to pick string beans before they are mature, when the seed is one-third grown or about one-fourth inch in diameter. If you get too many and there is not an organic food store that will take the extras, freeze them in late afternoon, preferably whole and as soon as picked. In warm weather, pick every day; when cooler, every few days. By frequent picking you make the season for each plant last longer. If all the beans on a plant mature at once, the bush gives up and dies.

This is why you get only one picking of the French Horticultural variety, those nice pinkish ones we call shell beans, which do mature all at once. I usually let the last picking of string beans go to full maturity and save them for dried beans to make soup or to sprout, though soy or mung beans are really better for sprouts. The pods will get brittle right on the vine, and also, sometimes, they get mildew. Therefore pick them before they are fully dry and spread them out on papers in the sun. Never let them get damp. A steam bath of three minutes will help to control weevils. Dry them quickly afterward.

Various pests and diseases might bother your beans, especially if they are not growing in soil that has been brought up to high fertility by humus and manuring. One is bean rust, which will respond to a dusting of garlic powder or a spray of water and garlic juice. This is also good for bean anthracnose and bacterial bean blight. The force of the garlic antibiotic is such that the University of California scientists use 1 part garlic to 20 of water, though this is much stronger than necessary, they say.

Try never to touch beans when wet; they bruise very

easily, and diseases move right in on the wounded spots. Organic material to put on or in the soil to help prevent fungus attacks includes: oat straw, mature soybean hay, and corn stover. Scientists at Beltsville, Maryland, have discovered that the streptomycetes which exude the antibiotics are especially attracted to the area by these particular organic materials. Such discoveries make the organic gardener who grows beans (and whatever else he wants to mulch with those straws) very happy, for they offer still another clear scientific reason to rely on the processes of nature instead of hard chemicals.

An old-fashioned cure for unwanted fungi, well understood before the days when man discovered antibiotics, is wood ashes. If mixed with lime and applied the minute a pest attack begins, they help.

Aphids can bother you, especially in dry weather. Get out the hose and turn it on the plants—undersides of the leaves if you can manage. Then use an onion-garlic spray on them. Plant nasturtiums among beans to control aphids. Strips of shiny aluminum foil on the ground beside the rows also will befuddle aphids, and the strips do an extra job of reflecting and thus add to the light the beans get.

If you get white flies and the brown curly leaves they cause, tear up the attacked bean bush immediately. Clean out all weeds such as mustard that might harbor them, and keep feeding compost. Garlic sprays help and pungent plants like tansy, mint, and wild marjoram. Use yarrow or the flower garden version, achillea, if you have no tansy.

Mexican bean beetles (brown, one-third inch, 16 spots) can be annoying if they go for your bean buds and young pods. One thing to look forward to is that they won't attack your late beans, anyway. Go after these beetles every morning and pick them off by hand. Also pick off their egg clusters on the undersides of the leaves. If the beetles get ahead of you, pull the plants up and bury them in your compost heap. Plant summer savory or nasturtiums, or move in some nasturtiums from a growing bed. Keep praying mantises, whose egg cases you can send for in the spring. Some have reported that planting potatoes nearby has a magical effect of repelling bean beetles. Easiest of all is to go back to the old stand-by garlic, and add a clove in each hill.

If you live in the South, you may have trouble from nematodes. For beans and all other garden plants, nematode enemies are controlled by marigolds—not only in the current year, but also for one or two years afterward. This means that you'll want to plant marigolds both for this year, where your sensitive plants will be, and also for the following year, where they will be in the future. Plow or dig them under.

One gardener reported that his best beans were planted right in straw mulch, spaced three feet apart. He happened to plant cucumbers between, and that is a good suggestion for anyone. You not only get their antibiotic effects to go along with that of the marigolds; you also get shading and sun at the right periods and provide maximum conditions for both plants through this arranged symbiosis. This gardener used no nitrogen fertilizer but nevertheless produced 200 bushels per acre of beans, with five pickings.

If you find you need a new recipe for string beans, heat 3 tablespoons of soy or olive oil in a frying pan and add 1/4 cup capers, 1 crushed clove of garlic, and 4 cups of previously boiled or steamed string beans, with 1/2 teaspoon salt and 1/4 teaspoon pepper. Toss them around until they are all well mixed and warm. Top all with chopped parsley and chopped spring onion.

But if all that seems too rich and highly flavored, try putting already cooked string beans in a casserole, sprinkling them with grated Swiss cheese, covering all with 1 to 1 1/4 cups béchamel sauce, and sprinkling the top with more grated Swiss cheese. Dot with butter, or soy sauce, and bake in a preheated 400° to 450° oven for 10 to 15 minutes. This makes a good vegetarian entrée.

Those two recipes disguise the flavor of string beans. Others who like the taste of beans suggest you add sliced mushrooms and butter, slivered almonds and butter, or juice of a half lemon and butter. All of these can be served with or without mixing in 1/4 cup of sour cream. Not bad. And not bad, either, as an addition to a green salad or a cold brown rice and pimiento salad.

Grow also pole beans, beans for drying, and lima beans (and see *Soybeans*). For heavy crops, and for saving space, grow pole beans. Each plant is big and needs space. Up on the pole or along the fence, where they are aired, they are less subject to ills. The old favorite variety, Kentucky Wonder, is still widely grown, and still notable for its deliciously beany flavor. If you let it go by and develop brown seeds, they are a good substitute for shell beans. The pods sometimes grow to 8 or 9 inches long. There is a rust-resistant Kentucky Wonder, and Scarlet Runner to grow on fences. Thin the beans to three plants per pole. All these mature in about 60 to 65 days. For a large wide Italian bean, try Romano, which matures in 70 days. They are stringless and tender. You can also grow pole lima beans, such as Prize-taker or King of the Garden, which take about 85 days to mature. They are quite vigorous, and very good for space savers. Try some at the side of a sunny patio, or along the front fence. Why not?

When the ground is warm (65 degrees for a spell of five days), plant Fordhook bush lima beans or Burpee's Improved bush limas, maturing in 70 or 75 days. For quicker maturing and for patio pots or window boxes, use Baby Henderson limas or Baby Fordhook bush lima beans. A packet will sow 15 feet, or five pots. They are somewhat fussy plants in cold climates, so it is best to start these beans indoors if you live where it is cold and damp. Set the seeds with the eye down, and see that they get moisture during their five-day germination period. Plant them in very good, well-drained, sodless soil, with a pH down to 5.5 (go easy on the liming and the nitrogen supply). Do not put any fresh fertilizer near the seeds when you plant them.

Easier to grow in cold areas, and substituted by many gardeners for lima beans, are fava beans or English broad beans. The variety to get is Long Pod, which will mature in 85 days and is much less fussy about the soil it germinates in. Plant early, for they do not like summer heat. The 7-inch pods are glossy green but not edible. Use as a shell bean.

Some people use an inoculant powder for beans. It costs only 40 cents or thereabouts, and will provide nitrogen-fixing bacteria. If you are gardening where sprays and chemical fertilizers have been heavily used, you may find this a useful aid to bringing back the organic liveliness of your soil. The small package treats 5 pounds of legume seed.

Beans for drying include navy beans, white kidney and red kidney beans. They are rich in protein, vitamins, and minerals, and should be part of any homesteader's garden. California red kidney beans take 100 days to mature, and one pound of seed will plant 150 feet of row.

The following recipe for vegetarian chili has been adapted from a soul food recipe. Soak 2 cups red kidney beans for one hour, then simmer for one hour. Sauté 2 chopped onions, 2 chopped cloves of garlic, and half a green pepper in 3 tablespoons oil until soft. Add half a hot pepper, 3 tablespoons very fresh chili powder, and 4 finely chopped ripe tomatoes. Simmer for 20 minutes. If you have 1 or 2 cups of water from vegetables, add at this time, and 1/3 cup honey, with 1/2 teaspoon oregano and 1 1/2 teaspoons cumin seeds. If no vegetable water is handy, make 1 to 2 cups in juicer—out of greens or carrots. Simmer for an hour and serve with corn bread.

beets

beta vulgaris

If you like beets at all, they are an easy, satisfactory, and nourishing crop to grow.

Varieties favored by experienced gardeners include Crosby's Early Wonder (55 days), Red Ball (60 days), and for late varieties Late Detroit Dark Red and Winter Keeper (80 days).

When you manage to get the seeds to germinate, to keep the early seedlings, and to transplant your thinnings instead of eating them, a 100-foot row of beets could yield you two bushels. This should be enough for a family of four with some to can and some to store in moist sand.

Plant one ounce for each 100 feet of row—or a packet for each 25 feet—and plant them early, for beets do not mind some early frost. Keep planting, at intervals of 5 to 10 days, until early July, so that you will always have young greens and small beets ready for harvest. See that the soil is in good soft condition.

Beets like a pH of 6.5, and enough nutrient matter and mulch for fast, vigorous growth. If it is slightly sandy, that is fine, for the roots like to have the earth fairly loose around them. Make the rows 12 to 20 inches apart if you are going to use a cultivator, but nearer together for mulch, such as salt hay, cocoa shells, or ground bark.

Warning: Do not plant the seeds too close together, because several plants come from one cluster of seeds. A good precaution is to soften and loosen these clusters after soaking the seeds for 24 hours before planting.

The flea beetle may turn up to bite the leaves of the young seedlings. Thin them if this happens, and spray with onion water mixture (one medium onion to a not quite full blender of water), and trust your vigorous plants to send out fresh, unbitten leaves. When the tomatoes and cauliflower come along, the flea beetle will probably move over to them, anyhow.

Keep the young plants weeded. When they reach about 5 inches, the tender greens are ready to thin and eat. In a week or so some of the roots, as you will see, will have widened enough to be small beets.

Beets provide vitamin A, riboflavin, folic acid, and

vitamin C. The greens are rich in this vitamin. Protect them by putting the greens in a dark bag as soon as you pick them, and into the pot within five minutes after picking, to preserve vitamin C.

In the fall, the rest of the crop can be dug and stored in a cold place (just above freezing). Though some people object to leaving beets in the ground until after the first frost, it really does them no harm. Select beets you plan to keep during the winter from rows planted later. In this way they will not be too heavy and coarse. In case your storage place tends to dry out, beets, like other root vegetables, can be stored in moist (but not wet) sand. Do not allow them to freeze. In the long run, canning may be the best mode of preservation because it's least risky.

For those who think they do not like beets, the first recipe to try is cream cheese balls with finely chopped beets rolled into them. The beets should be washed but not peeled and put through a meat grinder or blender just before mixing. Add finely chopped chervil, parsley, or tarragon for extra flavor. Rosemary and marjoram also go well with beets, but not the strong herbs such as thyme or oregano.

Another good variation for beets is a purée made by baking whole, unpeeled beets, then putting them through the potato masher. Add one-third their volume of thick cream sauce. Warm this mixture over low heat and add a tablespoon of butter or soy oil. Do not boil and do not stir. This can be sprinkled with a finely chopped mild herb and served in a mound or as a border around a pile of rice.

Another way to cook beets is to mince the tops and grate the beets. Then put these with a bay leaf and 1/4 teaspoon basil into a pressure cooker with enough stock to moisten the bottom of the pan—about 1/4 cup (a steamer can also be used). After quick cooking, add 1 teaspoon of honey and 1/4 to 1/2 cup yogurt. Odd, but good. What is left over can be put in a blender and then added to 1/2 cup milk powder, 3 tablespoons flour, and four eggs, separated and beaten to make a beet soufflé. Put in oiled casserole and bake for 30 minutes at 375°.

In late July when the beets in our garden begin to plump up, we like to sliver some to add to the salad, and also to braise. The slivers of four beets in hot vegetable oil to cover the bottom of the pan will cook very quickly. Then add 1 tablespoon of mixed water and lemon juice, and cook over moderate heat for 5 minutes. Put on the cover to keep the steam in. Make a mixed platter, sometimes, with carrots, peppers, and day lilies cooked the same way. You can use butter, but with delicate vegetables we like the milder oils like soy, peanut, or safflower.

beets . . .

(Some year, just for variety, grow white or golden beets. They are said to mature in 55 or 60 days, and to be exceptionally sweet.)

berries

If you have plenty of space for perennials, grow some berries. Try blackberries, boysenberries, dewberries, currants, gooseberries, raspberries (red or black, but not both), or strawberries. If you decide to attempt currants or gooseberries, be sure to see that there are no white pines in your neighborhood to be pestered by the fly that likes to spend part of its cycle in currants or gooseberries. Ask the county agent, for in many places there are state laws prohibiting growth of these berries. (Instead, think of elderberry, barberry, mountain ash, whortleberry, or Juneberry, also called shadberry or serviceberry—a delicious little fruit like an apple, wonderful after stewing to add to muffins, or make into jelly.)

blackberries, raspberries, dewberries

The brambles of blackberries, raspberries, and dewberries grow on canes that sprout up, bear fruit, and die back; therefore clear out old second-year canes as soon as the crop is picked. If you don't, you'll have a big messy bramble patch. New growth should be helped along by the gardener. Black and purple raspberries, dewberries, loganberries, and trailing blackberries all should have several of their tips turned under in three or four inches of earth in early fall. Keep the soil loose and moist, with plenty of organic nutrient, until the new roots form. Choose one-year-old plants for tip layering, and aim for four or five layerings from each plant. Be sure that you stick them in vertically, never horizontally. When you move the new plants, put them in deep, friable soil, three or four feet apart in rows six feet apart.

Red raspberries are suckered. That is, new shoots from the roots (or underground stems) are permitted to grow up to plants. After they get good roots, they are separated and replanted.

One of the pests on raspberries can be a mosaic that makes the leaves look spotty. Leafhoppers and plant lice bring in this difficulty, so try to get plants that are resistant to mosaic. These include Viking, St. Regis (in the South), Latham, Chief, and Van Fleet (in the West). If you do get

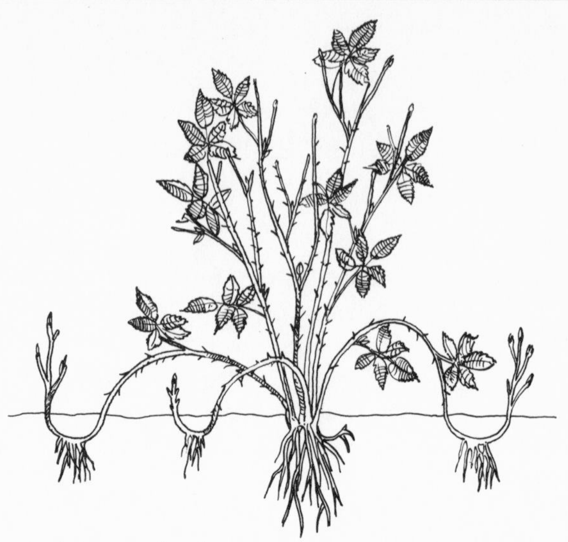

Layering Blackberries

Suckered Red Raspberries

mosaic, dig out and burn all infected plants. *Never* plant cultivated berries anywhere near wild ones. Of course, if you already have enough wild ones, you won't need to, anyhow.

Latham may be subject to gall too, a disease that makes knobs on the roots and canes. Destroy plants if you get it. Rust is another possibility and may turn up in anyone's garden. You know it by the red spots with white centers, on leaves or stem, and you can expect them to arrive after rain. Of all brambles, red raspberry is least likely to get rust, but as sure as I say that and get someone's hopes up, that is the very person who will probably have it turn up on his canes. The fungus causing this disease hibernates in infected canes, so try to see that everything you winter over is clean of it. Look for any cracks in the canes, or other signs it is there, such as berries that never ripened. Destroy all such canes. When you have to plant again, get only stock that is certified disease-free and move the field to an entirely different position. Once in a while you may see borers. The same treatment of ruthlessly cutting out all infected canes is recommended.

The descriptions of these diseases make it sound as though the growing of berries were a terrible chore and a terrible hazard. It is true that they are not exactly easy. But plenty of good sturdy mulch, patient attention, and a real enthusiasm for the excellent results you can get make the chore and hazard a lot less for organic gardeners than for some others. Thick leaves and manure or thick ground bark and manure would make a good mulch. A layer of folded newspapers at the bottom will add protection from weeds and confusion coming up from below to mess up your brambles.

One dish of berries for breakfast is a good start for the day, recommended in natural foods and yoga routines. Or you can make homemade berry juice in a blender or juicer, and this can be frozen and used later for sherbets, sauces, and jellies.

Make berry jam, with honey, adding 1½ cup of honey to 1 cup of pulp and letting it stand in sterilized jars for two weeks. (Honey has a powerful bactericidal property.)

Use it with homemade cottage cheese made from skim milk and buttermilk added lukewarm to a rennet tablet soaked in water. Heat this slowly and keep it at 110° for half an hour. Then let it stand for a day. Strain and wash out whey, then condition with ½ cup of cream or skim milk powder. For a gallon of skim milk you need only ¼ cup of buttermilk to get started.

If you prefer homemade jam with homemade yogurt, that is easily made by adding 3 tablespoons of commercial yogurt to a quart of milk. Heat the milk to lukewarm, add yogurt, stir, and cover with a warmed blanket. Let it stand overnight. A quart of yogurt dripped overnight through a double cheesecloth bag (or nylon stocking) makes cheese for breakfast.

I repeat: Do not plant until you find out from your county agent, your state department of agriculture, or an experiment station whether or not there is a quarantine in your district against these berries. If you are permitted to grow them you are lucky. Currants seem marvelously simple to cope with after the complications of the brambles. Almost any land will do, but they like cool, well-drained areas best. Add plenty of organic matter, such as leaf mold, manure, and hay mulch. Keep the plants pruned, and watch for the best fruits on the two-, three-, and four-year-old canes. Keep only the sturdy ones, and of course get rid of all that look diseased, cracked, or galled. Send away for new cuttings, or you yourself can take hardwood cuttings for new plants, and tie up 8-inch lengths to store in moist sand. By spring the ends will have hardened; plant them 4 to 6 inches deep and 8 to 12 inches apart, to grow roots. In the second year, move them to 4 feet apart. By the third year they will be ready to bear. Do not cultivate. Mulch. The delicate roots are too near the surface to hazard any instrument used for weeding.

If you see the foliage is curling and warping, you may have leaf lice. Look for its eggs in the bark of the plant's new growth. They hatch in early spring, so get to it early. Or late October and thereafter. Wash off what you can't pick off. Also pick off worms and scale bugs.

The rust disease is what gets into the white pine, where it is called blister rust. On the pines you'll see reddish spots on the trunk; on the currants the lower surface of the leaves get spots with curving tines coming out of them. Keep currants and white pine at least a mile apart.

All this applies also to gooseberries. But if you can have them, what marvelous tarts, jam, and gooseberry fool they make. For jam: Take 3 pounds of gooseberries and add 2½ pounds of sugar boiled 5 minutes in 2 cups of apricot juice. Add gooseberries and simmer for 35 minutes. Let stand for 24 hours before draining and putting berries in jars. Reduce liquid to thick syrup and cover berries. In a cup of gooseberries the food values include: 59 calories, 440 International Units of vitamin A, and 49 milligrams of vitamin C. Compared to these values you get from a cup of blackberries: 280 IU of vitamin A and 30 milligrams of vitamin C.

Of all the berries the one with most of vitamin C is the strawberry. In one cup there are 54 calories, 90 IU of vitamin A, and of vitamin C a high 89 milligrams. (In comparison: raw cabbage has 50; fresh cooked peas have 20; cooked spinach 50; though cooked turnip tops have 260 in one cup.)

To cultivate strawberries almost all soils in the United States are possible, but different treatments and varieties are required. The only soil not suitable is one that is very alkaline, though any dry soil will need watering, and a wet soil will need drainage. The best pH is between 5 and 6, a bit more acid than the soil of an average garden. Where it is cold try Dakota, Dunlap, Pocomoke, Crescent, and Howard 17; where slightly less cold: Howard 17 or Chesapeake, Gandy, Late Stevens, Dorsett, and Catskill (ask your county agent). For the Pacific, the *Organic Gardening Bulletin* suggests Clar, Dollar, Magoon, Marshall, Oregon, Wilson, and Nick Ohmer. For hot areas: Missionary, Aroma, Blakemore, and Southland. If you have a small family, the small crops from the everbearing kinds might appeal to you. Try Progressive, Gem, Green Mountain, and Mastodon.

Before planting strawberries, use your soil for something else for two years, cultivating carefully. Then set aside a portion of it for strawberries. The purpose of avoiding newly plowed land is to get rid of grubs which are always left over in soil just converted from grass. It will probably also get rid of wire worms, especially if you have planted marigolds during those two years.

A slope that has winter drainage is imperative, preferably facing south and protected from late frost. In the fall preceding planting add a good lot of manure (at the rate of 500 pounds per 1,000 square feet). Also dig in compost and/or leaf mold. Fifty or 60 bushels per 1,000 square feet would be a good dosage.

Buy plants and spread out the roots in deep holes half-filled with loose soft soil. Tamp this down a bit, add water, and fill the holes. While planting, avoid letting the roots get exposed to air or sun. Cover with wet peat moss if necessary. Keep the soil moist several days after planting.

We used to plant one spring, for harvest the following spring, pinching off all buds the first year to encourage green growth. We let the runners grow and developed walkways between every other row. You have to keep strawberries well watered, and well mulched especially in winter, with pine needles if you can get them. Fork the mulch over in the spring, and remove it for a while to help the ground warm up. Straw is a good mulch, too.

I haven't got around to trying a strawberry barrel for everbearing French berries, but it appeals to me because you fill the barrel with earth and grow the plants out the holes that are made in the sides for that purpose. Watering is easy if you run a hose down the center of the barrel. Of course if you run in liquid manure instead of straight water, all the better. Good loam and nicely spread-out roots are essentials. Put the barrel on a turntable so that all the plants get sun. Great for a terrace or patio.

elderberry

sambucus canadensis

A splendid bush to place around your lawn or in your hedge. It likes good soil and a sunny location, and will serve you well with blossoms in early summer to make a delectable tempura, and berries a month later for juice, jelly, and wine. In many parts of the country you will find it growing wild; it is hardy in the northern states. If you bring in cuttings and plant them, first in wet sand, then soil, you can get them to grow into bushes. Pest-free. Arabs used dried elderberry leaves for a repellent. You can also use the berries for lilac and deep purple dyes.

Juneberry

serviceberry, or shadberry: amelanchier

The beautiful white shad flower of early spring. Has small, dark-red fruits. Belongs to the rose family. The trees need some moisture and shade, and some species will grow 30 or more feet tall. Feed, water, and mulch well if you decide to get a few for yourselves or for the birds. Makes fine jelly, and is rich in vitamin C—as a cousin of rose hips should be.

broccoli

brassica oleracea italica

One year we had wonderful broccoli which we gathered right through the time of the early frosts, though at the end the heads were only the small side ones. But they were good. I hadn't yet learned the trick of cooking in milk and caraway to suppress the smell of *Brassica* family vegetables, so the next year, all were forbidden by the head of the house because the whole last half of the summer became associated in his mind with that smell. It is nothing but a sulfur compound anyhow, and broccoli ought to be worth it with or without milk or caraway for the excellent supply of vitamins A, B, and C as well as the calcium, potassium, and iron this vegetable provides. Pick while the head is still in bud and firm. After blooming, it gets flabby.

Plant your broccoli in the spring when the weather is still cool and moist, and keep well watered until it is established. If you live in a warm climate, you can also plant broccoli outdoors the fall before. Farther north the seed can be sown indoors or outdoors, but not later than the end of May. A March planting, indoors, if set out in May, might mature in late June. Give it some compost, but not too much. Broccoli does not need rich soil.

A packet will produce about 200 plants, if all come to maturity. That would take 200 feet of row. Green Comet is the best variety, but for a longer-growing follow-up, Waltham 29 is excellent, too. This variety has low plants, and fine, broad heads. (For plant care, see *Cabbage*.)

If you are a Gemini, broccoli is a food that appeals to you. People under this sign find a need for a wide variety of foods, to complement their many-faceted personalities. The following recipe for party broccoli is versatile enough to satisfy even a Gemini.

Sauté 2 tablespoons minced onion in a small amount of butter, or soy or safflower oil. Remove from heat, stirring in 1½ cups of sour cream or homemade yogurt (see p. 129). Add to this 2 teaspoons sugar, 1 teaspoon vinegar, ½ teaspoon poppy seed, ¼ teaspoon paprika, ¼ teaspoon salt, and a dash of cayenne pepper. Cook two heads of broccoli and arrange them on a heated platter. Pour the sour cream mixture over the broccoli and sprinkle with ⅓ cup of chopped cashews or ground-up dried soybeans.

Brussels sprouts

brassica oleracea gemmifera

Plants started in late spring in a cold frame may be planted out in early summer, spaced about 2½ feet apart in rows 2½ feet apart also, in well-composted ground. It is best to till in the compost two weeks before setting. Jade Cross is a vigorous hybrid variety which matures in 90 days. Start picking from the bottom, then remove the big lower leaves and keep right on picking until the snow flies. If you produce too many to eat, or to sell to your organic food store, Brussels sprouts are very easy to freeze for later use. They are also easy to transplant to a greenhouse or cold frame for late crops. They have to be kept moist.

Boiled or steamed Brussels sprouts can get boring, so dress them up once in a while with a sauce like hollandaise or spiced mayonnaise.

cabbage

brassica oleracea capitata

There is a tremendous advantage in growing your own cabbage because you can select and produce varieties that you never see in a market. Commercial growers choose to produce harder, more durable kinds, which unfortunately do not have the taste of varieties like Early Jersey Wakefield, Market Topper or Copenhagen Market, and Market Prize. The nutritional values of cabbage are excellent. There are in a half cup of raw cabbage 80 International Units of vitamin A (as compared to 20 in a baked potato) and 52 milligrams of vitamin C (about the same as a medium-sized orange). Not quite as much vitamin K as spinach, about 3.2 milligrams in cabbage, but of minerals you will get 46 milligrams of calcium, 31 of phosphorus, .5 of iron, and 7 to 24 micrograms of cobalt (40 in beet tops).

Since cabbage is a cool-weather plant, it is hardy and will tolerate a spread of temperatures. Give it plenty of compost, and some extra cottonseed meal or hoof and horn fertilizer for nitrogen. Get the soil into good loose tilth before planting, and try for a pH of 6 to 6.5. The early cabbages like the first two named do well if started indoors or in a cold frame, and set out after danger of frost is gone. Indoors six to eight seeds to the inch may be planted in a flat or Ferto-pots and lightly covered. The rows are best placed two inches apart. Harden the plants off by moving them to three inches apart when they are two or three inches tall and expose them to open air on good days.

Late cabbages, which can be kept for winter storage, are best planted outdoors in light but not too rich soil, on a date that will just about bring them to picking on the first day of frost and before they show any tendencies to burst. Growing times for different varieties are: Market Topper, 62 days; Market Prize, 76 days; Surehead, 93 days (flat and quite late); Premium Flat Dutch, 100 days (very large).

When spaced for final growth to maturity they should be set 12 inches apart for early cabbage, and 18 inches for larger ones. Protect the newly set-out plants with a basket, a brown paper shield, or even a shingle. Don't plant any specimens that are not strong and vigorously healthy. Keep the weeds down until you mulch, because cabbages have horizontal roots near the surface that need all the nutrients they can get. If you use a hoe, do it very gently so as not to injure the delicate rootlets.

At the first sign of a cutworm, get a tar paper or

other heavy collar to put around each stem, and watch every morning for intruders; also poke up any that lurk in the soil near the stem. A disk of tar paper laid on the soil around the stem will defeat the white maggot also, by preventing it from laying eggs near the stem just under the surface of the soil. If the leaves turn yellow and the veins black, you have something called black rot, and if your broccoli, Brussels sprouts, or cabbage leaves get black pimply spots, you have black leg. These diseases come in with the seed, but if you use the best heat-treated seed or if you buy plants from someone who will grow them to your specifications from pest-free seeds, you ought not to be bothered by such troubles. Good ventilation and a rich, well-composted soil will also help. Destroy all infected plants either by burning or by burying.

If cabbageworms turn up, spread wood ashes between the rows when the worms' white moths appear—the cabbage butterflies. But best of all, get out there with the tennis racket when they appear, and whack them. Broccoli and cauliflower are less likely to have these pests than cabbages.

In order to avoid club root, which is caused by nematodes, grow marigolds nearby—Mexican ones if you can get them. Also grow marigolds in the area of the garden where you plan to have cabbages the following year, and till them into the soil so they will exude the antibiotic substance which nematodes cannot endure. Burn up any infected plants. Do not bury.

When the cabbage head is large and hard, twist the stem in the ground, to break some of the roots, and thus stop further development of the head and avoid danger of cracking. On a dry, cool day a little later, harvest the cabbage and store it in a cold, dry place with a temperature of 33 to 40 degrees.

The best way to get healthy cabbage plants is to have good tilth, an airy location, proper spacing, good companionate planting, wood ashes, and habits of watchfulness.

Harvest directly into a dark bag, and keep the heads in a dark, cool place. Do not wash.

red cabbage

Is a delicious, sturdy variation. Grow both early and late varieties. The recommended early one is Red Acre (76 days) and for the late crop, either Red Danish or Mammoth Red Rock (97 and 100 days respectively).

Savoy cabbage

Is a favorite with a few people. In some places it can be sown in the fall. The heads are big and heavy, and one 90-day variety, Savoy King Hybrid, is likely to be heat-resistant. What people enjoy about Savoy cabbage is the mild, slightly sweetish flavor, the big handsome curly leaves, and the fact that it usually stays a nice green, when properly cooked. (No soda.)

Chinese cabbage

brassica pekinensis

Is an excellent cabbage for salads, and it can be used for braising, also. The Chihli or Michihli variety is considered a good one, or Wong Bok, which is shorter and stockier. One packet will sow 40 feet, enough for a family of four. It is good for freezing.

Since this vegetable will bolt, plant it in very rich soil around July 1 right where you want it to grow, for it does not like being moved. The root system is very sensitive, and needs a friable soil to stretch out in. Thin to 18 inches very carefully. Control leafhoppers with wood ashes or possibly rotenone. Keep well watered.

The entire plant is edible, though you may want to break off the outside leaves for tempura or boiling. Salad made from the inside leaves, if doctored with ginger as the Chinese do, can be crisp and tasty.

carrots

daucus carota sativas

By now most people like carrots, and if you are a Libra, you certainly do. What's more, if you are a Libra, you are a careful economist on food. Therefore it will pay you doubly, even quintuply, to reeducate yourself on how to grow, prepare, and take care of the good supply of nutrients which carrots can supply.

Saucy Carrots, described in the Libra chapter of Sydney Omarr and Mike Roy's *Cooking with Astrology,* is rather appealing. You start with butter, prepared mustard, honey, and chopped-up chives, warmed together before adding the carrots. You may prefer soy oil, well salted, to the butter—but salt all vegetables only after cooking to save nutrients. Simmer until carrots are done. It is better not to slice the carrots, and any reader, Libra or not, can see why after reading what J. I. Rodale says is lost through our usual maltreatment of carrots when we prepare and cook them. If the carrots are not fresh, 5 percent of the vitamin K and magnesium is already lost (especially if they were kept in the light). When the skin is taken off, there go another 10 percent of the nutrients. If they are sliced before boiling, all the vitamin C is lost, as well as the niacin and 20 percent of the thiamin. If you soak them along the way, all the B vitamins and some of the natural sugar plus all the minerals except calcium go out. (This would indicate that you should never put carrot sticks to soak in cold water to make them curl.) When the cooking water is drained off, down the sink goes all the rest of the vitamin K and more of the minerals. If sugar has been added to the cooking water, there is another 5 percent loss of nutrient, namely of calcium.

Last advice: eat your carrots raw, and by the way, eat the tops, too.

Shredding carrots just before eating them raw does increase the amount of carotene available to the body, by 5 to 35 percent. The tough cellulose walls of the cells are broken down, and the carotene released. But do so immediately before eating, as the loss in vitamin C will otherwise be great. In half a cup of carrots, the vitamin A supply is 10,000 to 12,000 units (twice the daily requirement), so, yes, they are indeed good for the eyes.

Carrots are best for you raw, but they taste so good in so many other ways, too, that you also want to try carrot chiffon pie, carrot-rhubarb jam, carrot stuffing, baked carrot ring, soufflé, and a half dozen other goodies. In her *Natural Foods Cook Book* Beatrice Trum Hunter gives recipes for all these. A vegetarian neighbor of mine makes a mixture of grated carrot, potato, an egg, and chopped chives, which she fries quickly in small patties in safflower oil. Delicious.

In the early days of this country the settler Francis Higginson wrote home to say: "Our turnips, Parsnips and Carrots are here both bigger and sweeter than is ordinarily to be found in England." This was written in 1630 and it reflects the fact that carrots grown in deep, rich, humusy soil, as the new soil then was in the Northeast, will be large and sweet and full of good nutrients.

In today's soil, this may mean you need big doses of compost, fall and spring, and a seedbed very well prepared. The soil should be broken up and the particles fine, for the seeds of carrots are tiny. In fact, they are so tiny that it is a good idea to mix them with the larger seeds of radishes and sow both together. Radishes come up very quickly, so you'll know where the row is and be able to watch for the very small seedlings of carrots when their lacy leaves first appear and will need some weeding; also the carrot plants will be farther apart and the thinning will be easier. Do it carefully to avoid disturbing the roots. In fact you may have to cut instead of pulling unless the soil is very loose.

Here is one warning. Your temptation will be to pull out the larger plants which may look just about ready to eat as tiny fingerlings. Resist this. You may pull one or two just because they are so marvelously tender and sweet to eat raw then and there at this early stage. But remember that the very small, weak plants are more than likely to be plants that won't ever amount to much. Pull these weak ones out and eliminate them from the contests for nutrients. They will be recycled, anyhow, when you put them on the compost heap. Plant carrots several times to keep these crops of young ones coming. Small whole carrots, quickly washed, blanched, and frozen, are a treasure to have in your freezer.

The spacing for carrots left in the ground to mature in the late summer should be no nearer than an inch and a half apart or they'll curl around each other. Your harvesting should be with the forethought that big strong carrots (Chantenay, especially) keep well in the ground not only until after frost, but throughout the whole winter under deep mulch.

To keep them in the ground over the winter, pile up autumn leaves over the row. If there is not a rain soon after you do this, weight them down so the leaves won't blow. Considering how hard it is to find plants under the snow, it is also a good idea to put some bushel baskets along on top, and more leaves over them to help keep out the cold. Then, when you want to go to the garden to dig carrots in the winter, you'll know where they are. The leaves will have prevented them from freezing. Carrots kept this way are very crisp and sweet. In spring, if the frost gets in again after a warm spell, the carrots will get mushy very easily. Some people lose their carrots during the winter from a period of alternate warm and cold that spoils the vegetable. Guard against this by adding leaves or sawdust to make the mulch airtight and manure to keep it warm.

If you don't think you'd have luck with this method, dig your carrots up and pack them in boxes of moist sand. The boxes can then either be stored in a cool place or be buried and covered with earth and then leaves, with some sort of marker for the spot. Do not use plastic—the carrots will rot like anything—and do not let them remain in the light.

With all these points in mind, select your varieties to suit your need. For earlier carrots to eat in summer, select Pioneer, the midget Short'n Sweet, and Scarlet Nantes, which are rather short and stumpy. The best later ones for storing are Royal or Red Cored Chantenay, and Commander, all very satisfactory, we find. The ones we winter in the ground are Chantenay.

Winter Protection for Carrots (Bushel Baskets and Autumn Leaves)

Do not be fooled by this *early* and *late* distinction. They all mature within a few days of each other if all are planted at the same time. These terms imply, rather, which carrots are considered appropriate for early or later plantings. The very earliest and latest plantings, however, are never as good as the middle ones, say between May 15 and June 15 in moderate zones. If you have heavy, clayey soil, plant Commander; if you live where you have droughts, choose a long variety with a long root, for example Danvers. Do not be surprised if you need a long spade to dig them out to eat; drought conditions make the root go way down into the earth.

Avoid using much manure, but use compost. And mulch.

One cool evening try Carrot Tempura: Have ready a batter; at the last minute cut the carrots diagonally very thin, and after coating the slivers in batter, drop into 350° corn oil, sesame seed oil, or safflower oil, two or three inches deep, and allow the carrots to come to the top. Turn and brown the other side. Batter: 1 cup pastry flour, 1 to $1\frac{1}{4}$ cups water, $\frac{1}{4}$ teaspoon salt. Mix lightly. Lumps don't matter. Variation: 1 cup whole wheat flour (freshly ground), 1 teaspoon cornstarch, $1\frac{1}{4}$ cups water, $\frac{1}{2}$ teaspoon salt. An egg can be added, if desired. This is adapted from an Ohsawa recipe. Decorate with tempura watercress, onion, thinly sliced burdock root, lotus root, whole radish leaves, or elderberry blossoms.

cauliflower

brassica oleracea botrytis

Go easy ordering seeds for this vegetable, because a packet will yield 150 plants, and for best nutrition and taste the head should be eaten as soon as picked; or frozen for use a little later. For varieties, choose Snowball or Snow King for white; and Royal Purple or Purple Head for the dark ones (which do not cook purple, but green). Quick growers are Snow King Hybrid, Early Snowball, and Snowball Imperial. Start seeds in the cold frame and plant in succession so they won't all mature at once. They are a delicate vegetable, or rather flower, and you need to keep watch for cutworms, cabbage loopers, maggots, and club root. Dust freely with wood ashes, and some of your troubles are under control. Have a very good soil, rich and loose, with plenty of

nitrogenous materials such as blood meal or cottonseed meal. These plants can endure cold, so can be put out in the garden quite early.

Place the plants 2 feet apart, or 15 inches if small, but discard all that have stunted leaves, for you need the leaves later to tie over the head as soon as it appears. Cut early, or when the head is about 6 inches wide and florets are still compact and tight.

A good steamed cauliflower under a slightly cheesy sauce is a very fine vegetable. Served in this way, it is usually left whole. Undercooked florets chilled with a sauce are second best. For the sauce: melt one tablespoon of butter and add 1 tablespoon of flour and 1 tablespoon of homemade mustard. (See pp. 178–9.) Mix this roux and add $\frac{2}{3}$ cup boiling water and stir. When melded, add the yolks of two eggs and

1 tablespoon cream. Beat well and cool before adding to the cauliflower florets. A very delicate hors d'oeuvre. Uncooked florets with a green dip of mayonnaise, chopped greens, and cream cheese is less delicate, but even better.

celeriac

apium graveolens rapaceum

Also called celery root, turnip-rooted celery, or knob celery.

Why this is not popular in the United States is hard to understand. It is quite easy to grow, and it produces nice large roots which are eaten any time after they reach a size of two inches. No blanching is needed, and the roots last into the fall and winter. What's more it is easy to store them, and to enjoy their rich, nutty flavor cubed, steamed, boiled with or without a sauce, in or not in a stew, or cooled for salads throughout the year.

It takes a long time for celeriac to grow, so choose the variety Alabaster and get started planting early. This is a thick, firm-rooted variety which takes 120 days and then may be stored for the winter in a cold cellar.

One recipe for celeriac which you may not have thought of is a purée of celery root and potatoes. Take about a pound of roots, wash, and take off any tough skin or fibers. Then boil until tender, and boil ¾ pound of potatoes. Mash together the potatoes, the celery root, and one hard-boiled egg with 3 tablespoons butter, ½ teaspoon salt, and a pinch of pepper. Reheat over a mat until hot again. Do not let it burn. You can also let this cool and mound it under mayonnaise for a salad. Either cold or hot, it is a welcome variation from the same old mashed potatoes, potato salad, or plain boiled celeriac.

celery

apium graveolens

Quite finicky to grow, and we never have, but those who have give good advice after their experiences. The pests are few, they say, but you have to have just the right kind of soil, moisture, and guards for the celery stalks so that they do not get too tough.

Compost and rich topsoil are obligatory, but the organic gardener will already have such soil for all his vegetables. Start them indoors or in a greenhouse, rather early, or buy the plants from a nurseryman. This means an eight-week preparatory period before the plants are ready to put in the ground. But they will withstand cold, so can be put out when it is still rather chilly. If the young plants, when you first set them out, need to be protected from hot sun *celery . . .* to make them sturdy, provide shade. Two thicknesses of material are best, with one removed after the young plant gets used to the outdoor environment. Do not plant where there has been lettuce or cabbage in recent years. Remove all diseased plants if spots come on them, or if nematodes get at the roots and cause galls.

When stalks are nearing the size you want, to make them tender and to keep the hearts growing, they can be blanched or banked with boards, held up by stakes; see illustration, p. 140.

Pascal varieties do not demand blanching. Anyone concerned with nutrition prefers pascal because of the added benefit of the green cells. Giant Pascal and Burpee's Fordhook are recommended, and also Harris's Tall Green Light, which resists the disease called yellows. One biological control of pests that get on quite a few vegetables, including celery, is a bacillus, called *Bacillus thüringiensis,* with the trade name Thuricide or Biotrol. Not everything is understood about how it works, but it is known to invade the larvae of insects at a stage when they are geared to eat continuously. This bacillus paralyzes the gut, and the insects cannot eat. Yellows is also controlled by planting clean seed in clean soil, and using such resistant varieties as Michigan Golden, Florida Golden, and Forbes Golden Plume. To fight blight choose Emerson Pascal. If you have nematodes, keep planting and plowing in marigolds. A sign of these pests is a browning of the rootlets and chlorosis or whitening of the leaves when the infection is bad. Leave the area fallow for two or three years, and always keep a celery bed absolutely free of weeds and trash. Celery can be stored in moist sand in a root cellar in the winter. It keeps well for several months.

chicory

cichorium intybus

Called Belgian or French endive but also (as in many seed catalogs) witloof (white leaf) chicory. It is a delicious tidbit to have in the cellar or dark closet during the late winter, to cut when practically nothing else freshly grown is available. Its cousin, Magdeburg chicory, is used to flavor coffee, and has gone wild along our roadsides—the sturdy weed with fragile, round blue flowers. In a good soil, sow the seed thinly in May or June, cover lightly, and firm down. When plants come up, thin to six inches apart; if you wait until they are three or four inches high, you can use some for greens in the summer. Let the rest grow. One packet of seeds should give you a final crop of about 25 feet. You will have no pests on it.

In the late fall, cut off the tops within an inch of the crown, lift the roots with a fork, and store in a cool place. Bring eight or ten into your cellar or dark closet, and plant them upright in damp sand or soil, covered with inverted boxes or something to keep out all light. You can also plant them in a box turned sideways between layers of sand, soil, or peat moss, with the box open on the side away from the wall. Then when they grow, and turn upward on reaching the air, they are open-leaved instead of in a tight thin head. Like carrots and parsnips, chicory can be left in the ground, if well protected by leaves, hay, and manure covered with a ridge of soil. Manure as the last layer will help to keep the pile unfrozen until the snow provides the final mantle.

Cut when ready, wash gently, toss in the finest French dressing you can make, or stuff with mashed avocado and a few drops of garlic juice.

I think that the best cooked chicory is braised. Melt 1/4 cup butter in a pan that has a tight cover. Put in 6 heads of chicory, 1/2 teaspoon sugar, and juice of half a lemon. Heat very slowly over a low flame. Push down a piece of well-buttered waxed paper over the chicory and quickly put on the lid. Reduce to very low heat for five minutes, then raise it a little for 15 to 20 minutes longer. Shake the pan once in a while to prevent sticking, but do not raise the lid. Add salt at the last minute and serve on hot toast.

collards

brassica oleracea acephala

This member of the cabbage family is an old favorite, eaten by Europeans ever since the days of Pliny, who named it *caulis,* or stalk plants. Our modern words cole, kale, kohl (rabi), and cauliflower (and the French *chou*) all come from that name.

The stalks and leaves of all members of the *Brassica* family, when young, make good greens, though Americans make use of few of them except collard leaves and, of course, cabbage. Collards are often grown in the South, because the plant withstands heat very well. Easy to grow, especially if you plant the variety called Georgia or Southern. In the South they can be sown both spring and fall, and left after thinning about two feet apart. In the North people plant collards in the summer, and let them mature for late greens. You can pick and pick the outside leaves as the plant grows. Take *collards . . .* them while still young and not yet tough. As the plants are stripped of their lower leaves, it is best to stake them. If you see cabbageworms, or other cabbage pests, treat them as you would on cabbage.

corn (for humans, is known as sweet corn)

zea mays saccharata

That Latin name shows why Europeans call it *maize,* and since this plant, for us and for the Indians before us, was a staple grain, we use the English word for grain, which is *corn.* It seems to express the early settlers' first surprise that the native plant they found was good enough to be a reliable crop in the virgin soil of those days.

Actually, what we call sweet corn was developed by hybridizers during the past century and a half. Maybe that's why it is subject to so many pests. With a small garden you may have to experiment to see which variety of this now temperamental and susceptible vegetable is best suited to your control program, climate, slope of land, and soil conditions. The hope of the plant breeders has been, of course, to bring out varieties that are high-yielding, vigorous, and more or less resistant to disease. When they happen to taste good, all the better. Long inbreeding and crossbreeding are now at such a high point of scientific exactitude that you can send away for prepackaged collections which will give a crop lasting over weeks instead of days and even provide many varieties of taste, color, and texture. In general, corn is white or yellow, but today you can even get Butter and Sugar or Honey and Cream, with kernels that are both yellow and white, as the names imply. They have fairly good taste and tender kernels, but I don't think they can replace the excellence of Seneca Chief or Wonderful. For the earliest corn, Royal Crest is recommended. Do not be surprised if its ears are small; they are meant to be. Country Gentleman is the old-time corn favorite, still obtainable from Burpee's. If you are short on space, grow midget varieties, for instance White Midget or Golden Midget, which yield ears less than 6 inches long, in a crop that matures in 65 days. Stokes has one called Polar Vee that matures in 50 days. These little corns are good for boxes and patio pots if very well nourished and well watered.

Wait to plant until all danger of frost has passed, though you can start a few seeds indoors or plant some in a big permanent cold frame whose top you remove when it gets warm, or even under a series of curved plastic guards or cloches which you can remove later. Warm up the soil before planting, and keep it warm. Have very rich soil, especially rich in nitrogen, for corn needs a great deal. Fertilize again when plants are two-thirds grown. We sow in hills, four seeds to a group, one inch deep, and with the groups two feet apart. Separate the seeds in each hill about two to three inches. Plant in blocks of at least four rows, not in one long single row, so that pollination will be adequate. If the silk doesn't receive enough pollen, the ears will not get fully developed kernels. A big packet of seed will usually plant about four 25-foot rows and yield 60 to 75 ears.

Make successive plantings to stretch the season, with the last planting one of the early-maturing varieties such as Royal Crest, Earliking, Spring Gold, or Early Sunglow. Some houses send bright pink seeds, dipped in an organic compound that is a fungus repellent. They look awful, but aren't, though there may be some mercury in the dip.

Watch the young plants coming up, and if you see weaklings, thin them out. Cultivate often and carefully until the plants are well established and you can put on the mulch. Do not use deep cultivation, for it will hurt the shallow roots and prop roots. Corn takes gallons of water and needs all the big strong roots it can get.

The corn earworm (greenish and striped) is also the tomato fruitworm, so do not plant the two crops near each other—that just encourages the pest. Predators to control this worm are: the polistes wasp; the ichneumon fly, which feeds on and lays eggs in the worm; and the trichogramma wasp (see Appendix), which destroys eggs—a good thing because corn earworms emerge three from each egg. Best of all, this wasp attacks early before the plant begins to succumb to attack.

You can also use a quarter of a teaspoon of mineral oil with or without a dab of pyrethrum applied to the tip of the ear of corn when the silk begins to get brown at the end. Ryania or sabadilla as a dust can also be applied at the tip, and you can set black light traps to catch the moths before they lay eggs at the tips of the ears. The larvae of earworms feed downwards, so if you can get them at the top of the ear before they begin their damage, you will have control. See seed catalogs for mention of varieties with a very tight husk (for example, Burpee's Honeycross), which means a strain that would be fairly resistant to the corn earworm. One more control is the woodpecker. Use suet to keep him coming to your garden all year round.

The European corn borer, a common pest, is the

larva of a night-flying moth, so in egg-laying time in May, when the moths are moving in on the leaves of all kinds of plants, trap them with black light lanterns. Seven weeks later they start a new cycle, and this time they lay on corn. If the larva survives to emerge, it will bore into the plant anywhere, into the stalks, the ears, or stems. The best defense is a thorough cleaning-up every fall, preferably by plowing under the infected stalks so they are buried, or by tilling them into the earth deeply enough to stifle the eggs. Other standard controls include the use of organic dusts or sprays such as rotenone, ryania, or sabadilla. Sprinkle the dust in the axils of the leaves where they emerge from the stem. Do it after a rain. If you can find and remove all the eggs, you won't have to spray, but it is hard to detect the scalelike white masses sometimes. The borer itself is easy to find. Remove it.

Though the ladybug prefers aphids, that predator will also control corn borers. Send for a box of them, open it, and let them out according to directions. The parasite *Lydella stabulans grisescens,* a fly, is sometimes used in the eastern and middle Atlantic states and the spores of the fungus *Beauveria bassiana* are found to be effective against newly hatched larvae. Also the easily obtainable *Bacillus thüringiensis* works effectively.

The nastiest pest on corn is a fungus called smut that causes first a hideous white bulge which is a gall, and then, when the spores burst out, a mass of ugly purplish black. Take out and burn any galls you see, but do not rely on your compost heap to consume them. If the fungus happens to survive, it might infect your whole pile. Burpee's Honeycross is said to be resistant to this as well as to earworm and wilt. Order that variety.

Again, good soil is the best preventive of all. One correspondent quoted in *The Organic Way to Plant Protection* says, "Several times I have planted a block of corn in my best soil and continued the rows into poorer ground. Borers and earworms have invariably attacked the unnourished plants, while the rest were left strictly alone. I am sure that rich organic soil is the answer to many common problems of plant pests and diseases."

Another pest you can get is raccoons. They absolutely love corn. Wrap up each ear in both a paper bag and a wire net or screening. I hope you have an army of people to help do that tedious job. In spring crows, which do not always cause trouble, sometimes try to dig under the very young plants for the remaining part of the seed kernel you planted. The best way to foil them is to use a good, deep mulch until the little plants have used up the stored food in the endosperm of the seeds before the crows see them.

corn . . .

When you get ready to eat corn, remember just this: the sugar begins to turn to starch twelve minutes after picking. Have the water boiling or the pressure cooker heating before you even go out to pick. Cook five minutes or less in unsalted boiling water. All stale corn is second rate.

Roast corn: a delicious way to prepare corn. Pull down the husks, but do not pull them off. Remove the silk, and butter or oil the kernels and salt them if you wish. Pull the husks back up, tie them, and roast over coals for 15 to 20 minutes or in a 350° oven for 30 minutes or until tender, turning them often.

For freezing corn, be sure you get the ears into the blanching water while there is still sugar in the kernels. Eat frozen corn within a few weeks. There is loss of taste and sugar if you don't. A second method for freezing is to tie up ears, husks and all, put in individual plastic bags, and freeze without blanching. Again, the corn should be absolutely fresh-picked.

If you have a flour mill, dry all your leftover corn and have it ready to make into fresh-ground corn meal for hoecakes and hush puppies. Keep watch while it is ripening, for then is the time raccoons are especially eager to get at it. After harvesting, put it up where they cannot get at it. And save some corn meal after grinding for a soap powder for cleaning dishes and pots. It feels soft and smooth on your hands.

cover crop

It may be that you have inherited some old worn-out or previously pesticided fields which you want to renovate. One thing to do is to plow up the field and then use a cover crop, which is a crop you grow for the express purpose of plowing it under as green manure. *In the summer:* buckwheat, soybeans, millet, oats, and Sudan grass are suitable, sown in May, though in some areas oats may be sown earlier. Most of them can be plowed under in August. If there is a good yield, the crop should be cut and chopped before it is turned into the soil. Soybeans are excellent for a legume crop, but harvest the beans before plowing them under. (See p. 195.) Other legumes to consider include: common vetch, purple vetch, bitter vetch, horse beans, Tangier peas, field peas, *cover crop . . .* fenugreek, and bur clover, depending on your climate. Those with heaviest growth would probably be purple vetch, hairy vetch, horse beans, and Tangier peas. *In the winter:* winter rye is the favorite for a cover crop, and it will also protect bare soil from eroding. This and winter wheat can be planted in most areas between August 15 and September 15 at the rate of 2½ pounds per 100 square feet, and plowed under the following April. The fertilizing power of these crops will be about 2 percent nitrogen, .5 to .8 percent phosphorus, and 3 to 6 percent potassium.

Cowpeas are well-rooted legumes valuable where heavy crops have depleted the soil. They make sandy soils more compact and clays more friable because they add humus and regenerate the bacteria.

For your own table, soak some cowpeas for several hours in plain water, then simmer over low heat until tender. Add onion and garlic, pork fat or bacon fat, celery and green pepper. Recommended soul food.

cress (peppergrass)

lepidium sativum

Garden cress is not as good as watercress, but if you begin every dinner with a salad, the way Adelle Davis advises you to, you want some variety. Burpee's Curlycress can be grown the year round (in 10 to 20 days), indoors and out. Slow to bolt, too. Excellent for salads and green drinks. Less curly, but also up and edible in 10 or so days is Salad Cress from Burpee. Mild. Wonderful for sandwiches. It is very easy to grow, and practically pestless.

cucumber

cucumis sativus

We have grown cucumbers in rows, in hills, in the open, in the shade of corn, on and off compost heaps, and have had everything from bad to extremely good luck with them. The worst was when we let the grass get in; the best was when the vines had the most compost and some shade part of the time. One packet of seed will grow twenty hills, and that's plenty if the vines are kept well picked. To prolong the season start some seeds in Jiffy pots, and set them out under Hotkaps so they will get used to the outdoors as soon as they can stand it.

If you have some big pots to put on a terrace, plant some cucumber vines in one or two. With a good rich soil, an adequate watering system (mostly just remembering to do it), and some sort of support to keep the vines from sprawling, you can have a handsome and convenient pot plant. If the plant description in the seed catalog mentions that the cucumber is black-spined, that means that it is a pickling cucumber and has little black slightly prickly knobs on it. A good all-purpose pickle cucumber is National Pickling, recommended for all sizes from very small to six inches long. A prolific small pickle variety is West India Gherkin, which matures in 58 days.

Any of these will do better on a field that has been used for beans, lupine, or clover the previous year, and on an area scattered the week before planting with at least four-month-old manure as well as compost. Cucumbers need lots of nutrients; if you plant in hills, thin the plants to three per hill, so there will be enough for all. They grow very fast; be sure to cultivate to make room for the fast-growing roots, unless the soil is very loose to start with. After the plants reach eighteen inches, add a high-nitrogen fertilizer. Water at once, or apply just before rain. If you choose varieties like Poinsett or Gemini, they will be resistant to such pests as mildew, scab, anthracnose, and even mosaic and angular leaf spot. But you will have to keep watch, even on well-manured, well-composted plots, for the striped cucumber beetle.

As soon as he arrives, dust plants with wood ashes, rock phosphate, or granite dust, or use a spray of wood ashes and hydrated lime diluted in two gallons of water. Some people dust with rotenone. (The same holds for squash, pumpkin, and melon vines.) Unhappily, these pests come when it is dry weather. Lots of water and good mulches help to keep in the moisture. If the dust doesn't work in dry weather and if you do have plenty of water to spare, give the plants a good dousing before applying the spray. One difficulty to remember about dusts is that the natural predators are sometimes discouraged by their use. Hosing often does the trick by itself.

Other measures to take would be companionate planting with onions and garlic, planting marigolds, and permitting companionate weeds such as lamb's-quarters and sow thistle to grow up. It also helps to plant beans, corn, and peas nearby. Black light trap lamps have been used with considerable success on cucumber beetles at some experiment stations. Also keep the garden free from debris.

When your cucumbers get long and filled out and green, pick four to make Breaded Cucumber Slices. Leave them unpeeled, but wash off dust if you have to. Then slice them ⅛ inch thick, rub each slice with ½ clove of garlic, *cucumber . . .* and dredge with a mixture of 3 tablespoons bread crumbs, 3 tablespoons flour, $\frac{3}{4}$ teaspoon salt, $\frac{1}{8}$ teaspoon pepper. Heat $\frac{1}{4}$ cup soybean or sunflower oil in a skillet and brown the slices on both sides. Drain on brown paper and serve right away as a hot vegetable.

If this is too rich for you, braise them or make soup or that cold cucumber aspic which is always made from a secret recipe. One I pried out of an English hostess starts with one tablespoon of unflavored gelatin, soaked in leftover spinach cooking water for five minutes. Then dissolve it over hot water and add 1 cup of green juice from blended salad greens flavored with 1 teaspoon salt or less, $\frac{1}{4}$ cup lemon juice or less, 1 tablespoon of chives, 3 teaspoons parsley, and a tablespoon of pepper, all blended together. Add $\frac{1}{2}$ teaspoon fresh-ground white pepper if wanted. Let this nearly set, then add $1\frac{1}{2}$ cups cucumber chopped fine and $\frac{1}{2}$ cup sour cream or yogurt. Serve on a bed of young mustard greens, watercress, and small kale leaves, with very thin slices of cucumber as decoration.

People no longer believe in soaking cucumbers in vinegar, nor do they think they have to salt and weight them or peel them if they are home-grown. Unfortunately the store-bought ones that have been paraffined for long shelf life do have to be peeled. That's one additive you know you do not want to eat.

Cooked cucumbers are somewhat bitter, but they are one variation to use during the season of cucumber surfeit, and are very easy to prepare. Peel, cube, and boil for 5 to 10 minutes, depending on the age of the vegetable.

dandelion

taraxacum officinale

Used everywhere from ancient times as a source for vitamins in early spring tonics, usually made from the roots; and gobbled up as greens by almost anyone in the rural areas who can get out to dig when the days begin to get warm. Some cook them, can then, store them up for the following winter; others eat them raw and relish them in salad. Suburbanites still fight them as lawn weeds, but more and more now eat them when young as well as eradicate them. The old plants are really bitter, and lawn dandelions are never as tasty as those that come up in the fields or vegetable garden. If you save the crown of buds, and cook it with the leaves, it has a smooth, bland taste that goes well with the stronger taste of the leaves. The roots can be easily peeled and sliced thin, boiled in two waters, and buttered and salted to eat like parsnips. To grow your own dandelions, plant in rows 18 inches apart and cover with ½ inch of good soil. Thin seedlings to stand a foot apart, and use the thinnings for greens. The roots can be used for a coffee substitute if kept to the second year, when they get quite big. Either year, they can be brought into the cellar and planted as you do the roots of witloof chicory (see p. 143).

If you want to keep on eating dandelions all summer, start with seeds bought from a seed company, such as the variety called Thick-Leaved. From 25 plants you can freeze or can a winter supply and have fresh greens besides. Condition the soil with compost, and apply plenty of wood ashes and rock phosphate, and fish emulsion in water at 2 or 3 percent. For tender hearts of dandelion, let the outside leaves grow and then tie them up for a week to blanch the inside leaves. They will make a delicate and mild addition to your salads.

Dandelions are also easy to grow in a window box or patio tub. They can be interspersed with parsley, ten-day cress, tarragon, and other rugged growers. Keep adding nutrients at three-week intervals because the contest between the roots will be considerable.

Dandelion Wine: In one gallon of water, boil 4 pounds sugar until dissolved. While still boiling, pour mixture over the peels of 2 oranges and 1 lemon, which you have placed in a crock. When liquid is lukewarm, add 1 gallon dandelion flowers, juices of the fruit, and 1 ounce yeast. Cover, and leave for 10 days. Stir daily and remove any flowers that go bad. Strain the liquid into a clean jar. Secure the cheesecloth over the jar and press down in the center, forming a funnel. Tie on two thicknesses of cotton cloth for the top. Keep the jar in a warm room. Maintain it this way as long as little bubbles continue to rise.

When it's quiet, begin clearing. Siphon off the yeast deposit. Cover tightly and set the jar in a cool place for a few weeks; then siphon off accumulated yeast again. Continue this until the wine is clear. This could take three months, but you'll be glad to know that this process improves the flavor. Never bottle too early; corks could fly and bottles burst, wasting the wine.

Use champagne bottles, if you can get them. The indentation allows for expansion. Sterilize the bottles in hot water with soda added. Rinse and dry in a warm oven. As you remove each bottle from the oven, plug with cotton. Place two raisins in each bottle before filling. Leave space for the cork to be pushed in until level with the top. Wire the corks (you can buy loops). Store bottles on their sides in a cool, dark place for six months.

day lilies

hemerocallis fulva

These hardy flowers, which grow almost anywhere, with practically no pests, are wild lilies in Europe and Asia. At several stages they are excellent food. The small stalks when they come up in the spring provide a delicious, nutritious substitute for asparagus. As can be seen in Japanese and Chinese cookbooks, the day lily is an important food in the East, eaten as root, stalk, bud, and opened flower. They are often served, as tempura, dipped in batter and then fried (see p. 139). I braise all parts of them in butter over a slow heat, and serve on toast for lunch, or even breakfast. If you

Edible Stages of the Day lily: Sprouts, Flower and Buds

live out in the country, you may know of a stretch of abandoned land along the roadside where the common browny-orange day lily has gone wild. The sprouts on fresh, firm roots from that patch, if they are available to you, are just as delicious in the spring as those in your own garden would be. The bud and flower are also excellent. By ordering and planting cultivated day lilies that bloom at different seasons, you can have a good crop coming on from June to September. Many nurseries have them for sale at fairly reasonable prices.

dill

See *Herbs,* p. 164.

eggplant

solanum melongena

If you have rich, somewhat sandy soil and can start seeds in a hot bed or indoors over an electric coil, you might have some luck growing eggplant. Start with seeds in early April or whenever for you is eight weeks before the safe time to set out small plants in the garden. Plant the seeds in small individual Ferto-pots or Ferto-pellets, then transfer them to bigger Ferto-pots which you later plant in the ground, pot and all. Do not let them wilt by day or get chilled at night. Do not plant them out where peppers or tomatoes were grown the previous year. Cover with a basket or Hotkap if necessary. A dusting with rotenone or wood ashes now and then when the plants are still young will help protect them. One packet has about 35 seeds, so you would get more eggplant than you'd need if they all germinated. Perhaps the best solution with this rather difficult plant is to try your luck with eggplants started by a professional at a greenhouse. Six or eight plants will be plenty for a small family. Set them in a one-foot hole, with a good shovelful of compost worked into the bottom. The recommended variety is Black Magic Hybrid with a maturing rate of 72 days after setting them out. An earlier one is Burpee's Early Beauty Hybrid, which matures in 62 days after setting out and is very prolific. After the soil is warm, mulch deeply to hold up the fruits. Pick all the fruits as soon as they are plump and glossy, with their seeds small. Your crop will last for several weeks. Do not let them get overripe and pulpy.

The results of growing this difficult plant can be so delicious that the struggle is worth it. The best recipe I know includes a cubed eggplant and a finely chopped onion, sautéed in butter until the eggplant and onion are soft. Then add salt, a tablespoon of chopped fresh basil, one or two beaten eggs, and a handful of grated cheese. Add these all at once unless the cheese you use won't melt easily, and stir only until the egg is thickened and the cheese soft and runny. It is impossible to give exact proportions because the size of the eggplant will vary. You have to practice to get the mixture soft and smooth, and not have lumpy egg or stringy cheese. When done just right, it is a good hearty vegetable entrée. Use a little white wine to keep things moist if the juice from the eggplant is not enough. Variations on this dish, somewhat less of a gamble, are a custard with onion and eggplant, and a yogurt, wheat germ, and eggplant sauce which can be used with lamb or soybeans.

endive See *Chicory,* p. 142.

garlic See *Herbs,* p. 163.

geraniums
pelargonium

Grow lots of them, and after you have one plant, you can easily make slips by cutting off lengths of stem with several nodes, leaving them exposed to the air overnight to form a callus at the place of the cut, then putting them to root in wet sand. After they are well rooted, transplant them to good potting soil. In summer plant geraniums around your garden, on the outer edge and near corn or other vegetables that attract Japanese beetles. White geraniums, especially, are repellent. Take up all plants in the fall to keep indoors over the winter. Make the slips in the spring or summer.

grapes
vitis vinifera

Have wild grape vines if you can; they are sturdy and except for Japanese beetles fairly pest-free; and they make wonderful jelly and conserves. Otherwise get some two-year-old Concord grape vines to start out with. These are fine blue grapes, excellent for grape juice and jelly. They stand the cold, as do the red grape Agawam and the white grape Niagara, which ripens about the same time as Concord. Buy sturdy stock from a reliable nurseryman, and plant the vines in a warm south-sloping place, against a building or fence but not near trees.

Put in 12- or 14-inch holes, of about 16 inches in diameter. Mix bone meal, compost, and granite dust with good loam for planting. Whether you plant in spring or fall, prune the tops in the spring to a single cane with two buds. Have several vines of each variety, and look forward to harvesting the second year. You might get up to twenty pounds per vine when they get established.

Sometimes it is necessary to put a net over the crop if the birds find the grapes before you are ready to pick. Grape leaves are very attractive to Japanese beetles. I have never found that they did much harm to our vines, but it is obvious that any deprivation of leaf area robs the plant of some of its food. Pick off the beetles and try garlic-onion-and-water spray if they persist.

We have not yet tried hardy seedless grapes, but they sound very attractive. Recommended varieties are Interlaken, New Himrod, and Pink Aurora, which is not really seedless. Interlaken is said to be good to combine with Concord, for it ripens three to four weeks earlier.

Any time after the leaves come out you can begin to get the benefit of rice, soy grits, onions, and rosemary wrapped in grape leaves and simmered in a casserole until the leaves are tender. We use a cup of cooked rice, ½ cup cooked grits, 1 tablespoon wheat germ, 1 tablespoon low-heat milk powder or soy milk powder, ½ cup chopped, sautéed onion, and 1 tablespoon crushed rosemary. Mix well and wrap in grape leaves, putting about 2 or 3 tablespoons of the mixture on each leaf, depending on the size. Pin with toothpicks. Use chicken broth or tomato juice for the liquid in the casserole. Bake at 325° for 40 minutes, or until leaves are tender. Also nibble some young tendrils. They have a sweet-sour taste.

herbs

On many occasions you will be glad that the methods of organic gardening demand plentiful crops of herbs. Many you will begin to clip to add to salads and other foods while still young and tender. Many you will plant here and there among your vegetables and on the outer borders as pest repellents. Many you will harvest on a warm, clear sunny day in summer when your herbs are in top condition just before blooming.

Prepare the soil as you would for vegetables, and let it settle a while before planting. Some seeds can be helped along by letting them soak on a saucer in the kitchen before planting. Others can be started in the cold frame. Some come up well in full sun; others need to be shaded when young. But if you follow the general rule of shielding young plants from the very hot sun, and from drying out, you will have pretty good luck.

If you are inclined to grow too many herbs, and too much of each, harvest them anyhow, and give them away or take them to the organic foods store. They also do well on the compost heap and, in fact, the Bio-Dynamic gardening group systematically incorporates camomile, nettle, and several others into each heap.

The "basic compost herbs" are: chicory, nettle,

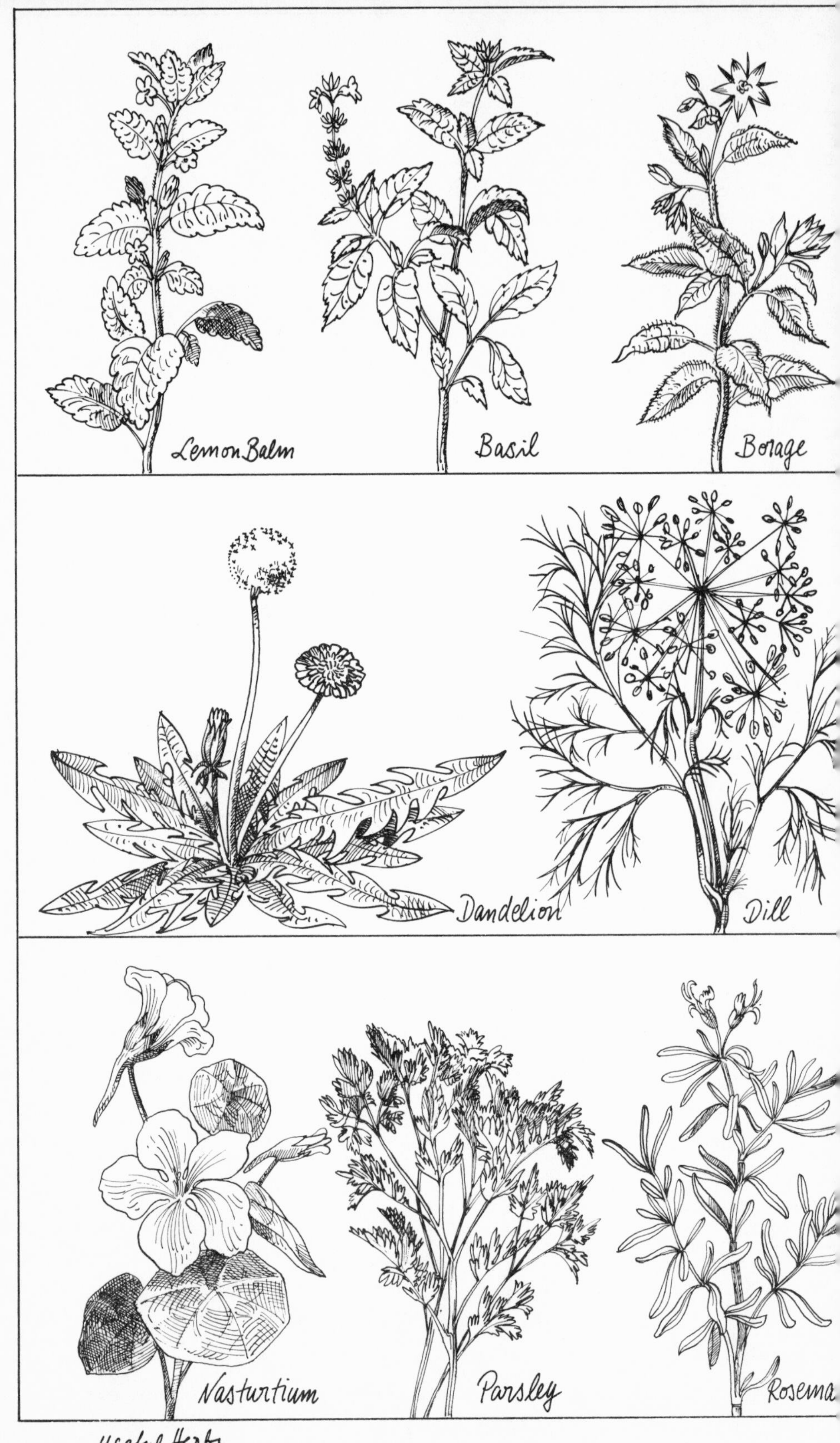

Useful Herbs

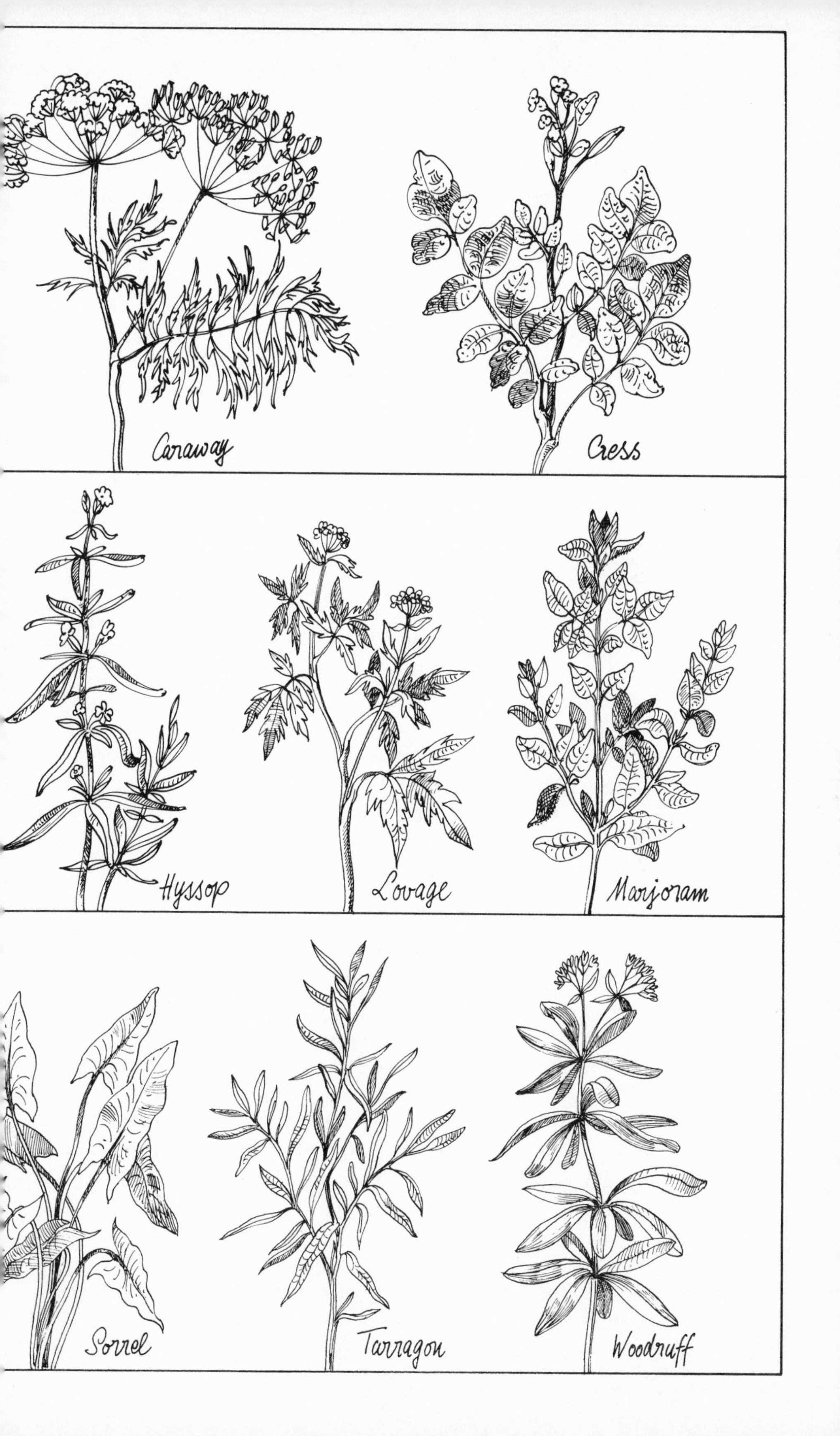
Caraway
Cress
Hyssop
Lovage
Marjoram
Sorrel
Tarragon
Woodruff

camomile, dandelion, yarrow, valerian and (though not an herb) the ground bark of oak trees. Specific virtues claimed for these herbs are: for chicory, an alkaline salt and silicic acid; for nettle, a sulfuric content that regulates potassium, calcium, and iron; for camomile, a powder to mediate between the silicic acid of the cosmos and of the earth; for yarrow, a guiding power for potassium because of its affinity for sulfur and an ability to draw substances into the soil and the compost heap from the atmosphere; for valerian, a powder to help phosphorus compounds be assimilated; for oak, to aid the health of plants by its calcium content.

Pests are rare on herbs, but if you get red spider mites or aphids, wash the plants with soap and water or hose them down; if they are attacked in the home apply tobacco, rotenone, or pyrethrum. But plants that are sturdy and well nourished probably won't get such pests.

In planning, choose both pest-repellant and culinary herbs, such as basil, chervil, sweet marjoram, thyme, rosemary, and tarragon, and the old stand-bys chives, parsley, summer savory, and dill.

Remember that rosemary, the thymes, sage, and winter savory are quite pungent, and slightly less so are the mints, basil, tarragon, and sweet marjoram. Not usually eaten but strongly pest-repellent are feverfew, tansy, marigolds, and nasturtiums, though I have a recipe for marigold custard and I often put nasturtium seeds and blossoms in salads. Some perennial herbs, such as chives, thyme, rosemary, sage, and lemon balm, can be potted and kept on the windowsill or under lights.

Perennials outdoors will also be ready to use the first thing in the spring after the ground thaws and the new shoots come up. A semihardy plant like marjoram can also be planted in the fall if you live in a mild zone.

When you harvest herbs for drying, cut only the top of the plant, about the top third or quarter of its growth. (Parsley and chives can be cut farther down.) Then tie the herbs in small bunches, or put a rubber band around them, and hang them to dry in a cool, airy, darkish place. It is quick drying, in the shade, that preserves the flavor, texture, and color. If the weather suddenly gets muggy, transfer them to a very slow oven, about 200°. Watch them often, and take out as soon as they are dry, for they can lose their flavor if left too long. Then you might as well throw them away, unless you want to mix them in with other herbs with a strong flavor for herb teas. (Mint would be a good coverall.)

When done, strip the leaves from the stems, pack in small glass jars, and store.

Freezing herbs is less satisfactory than drying, but steeping them in white vinegar for two or three months is a good way to preserve the flavor. One of the best ways is to make various kinds of herb butters; they give a much better flavoring to vegetables than butter and herbs added separately at the last minute. The herby flavors have had time to be extracted and concentrated. If left undisturbed, a jar of herb butter in the refrigerator will keep six months. Once the jar is open, use it up for sauces, sandwich spreads, to flavor scrambled eggs or to rub on chops, steaks, hamburgers, or roasts.

angelica

angelica archangelica

Easy to grow in any good soil, perhaps in part shade, and perhaps in a slightly moist place. The tender leaf stalks can be chewed as a substitute for candy, put into cakes, or used as candied angelica for cake decoration. Let it resow itself, for it is a biennial. Nichols in Oregon (see Appendix) has seeds and plants.

anise

pimpinella anisum

A fragile annual with a licorice taste. If you sow it outside, do not thin or move it, for it cannot stand upset. Put it in the sun, and harvest it for the seed the minute the first seeds are ripe. Spread in a warm, dry, shady place on papers to dry out. Use for desserts and drinks.

basil (green or purple, called ornamental basil)

ocimum basilicum

Easy to grow in flats, but hard to start outdoors in the hot sun. If you plant it in the ground, see that the first shoots are well shaded and well watered. It is a native of India, and a very useful herb. It can be harvested all summer from the bottom leaves. Use the cut tops to dry out for winter use. Basil is a delicious addition to eggplant and sliced tomatoes, and indispensable in soups.

borage

borago officinalis

This annual herb, much loved by bees, is used as leaves or as a flower in drinks such as lemonade and Pim's Number One Cup. No wonder it has had a fine reputation for curing melancholy, reviving hypochondriacs, and cheering students. It can be found, if you want to hunt, around old dumps and abandoned houses, probably as an escape, for it reseeds *borage . . .* readily. The young, tender leaves have a cool cucumbery taste for salads, and it will add potassium and calcium to your diet (or to your compost heaps). Press out some juice for freezing, to add to jellies, wine bowls, and candies.

bouncing Bet (soapwort)

saponaria officinalis

Brought here by our forebears to help with the washing. The juice, says an old herbal, "scoureth almost as well as Sope." Old ladies with secret skills clean up and return the color to faded damasks and tapestries with the aid of this herb. It has also been used to scour one's inside, for example, "against the French Poxes" and "the hydropical waters." The gaily flowering bouncing Bet grows wild along the river in our town, and has been brought into several gardens to add its pale and darker pink blossoms to the summer display. It is easy to grow and charming to have around, but it does need watching because it spreads very quickly. It can be bought from many nurseries. Use leaves as a laundry aid. For soap tie ten stalks of bouncing Bet together in bags. Boil in soft water until a soapy froth appears and the water turns greenish. First soak the fabric in cold water, rinsing it until it is clear of dirt. Place the fabric on a board over the bath and work in the soapy froth, using a sponge and moving in a circular motion. Keep applying as long as there is any dirt at all. Finally, wipe foam away, and dab off with towels. Dry in well-ventilated, shady area. (See *Horsetail,* p. 166.)

burdock

arctium lappa

Though the burs of this big biennial plant are hideous to get in your hair, or in your dog's fur, the root, sliced and boiled, is very good, and its food content highly valued—especially by Ohsawa and his followers. If you want to grow it, you can find it on almost any vacant lot or along almost any old railroad. It is hard to dig, unless it is growing on very loose soil. In Japanese markets in this country the root is known as Wild Gobo. One of the best ways to cook it is to fry it as tempura after it has been boiled tender. In early spring, the young shoots are good, but you must peel them, for the rind is quite bitter. Just eat the inner pith.

caraway

carum carvi

An herb for seeds, to put in your breads and cakes and, in fact, to reproduce the Seed Cake which Molly Bloom eats on Bloomsday. The young leaves and shoots can be used for salads to give a strange tang and also to quell cabbage-cooking smells. Like parsley, this is a biennial, so count on planting seed every other year, or plant annually, in a permanent garden, in the same row, between the plants already there. Dress with compost.

caraway . . .

Molly Bloom's Seed Cake: 1 cup butter, 1 cup sugar, 4 eggs, 2½ cups sifted flour, ½ teaspoon baking powder, 2 tablespoons Irish whiskey, and 1½ teaspoons caraway seeds. Cream the butter and sugar until white and fluffy. Add the eggs one at a time with a dust of flour each time to prevent separating. Beat well after each addition. Fold in sifted flour and baking powder, add seeds and then the whiskey. Pour into an 8-inch cake pan, lined with wax paper. Scatter some caraway seeds on top and bake for 1 hour at 375°. Reduce the heat for the last quarter hour to prevent burning.

catnip

nepeta cataria

Why not? Lots of people, also, like it—for tea and as a seasoning. Eighty days to mature, and it grows 1½ to 2 feet tall and is perennial. Burpee has seeds. Not fussy about soil. Good pest repellent. Excellent for compost and green fertilizer.

comfrey

symphytum asperum

An herb found in many an organic garden, because of its nutritional values for man, beast, and soil. Send away to a specialty house for pieces of root, and plant them in good loam with a pH of 6 to 7, at about 6 inches deep. The excellent root system of comfrey taps and brings up many minerals. Its fast, heavy growth chokes out weeds, even when planted 3 feet apart. It should be cut 2 inches above the ground, about five times a season. In the late spring, when the shoots are tender, begin the first harvest for salads.

Since it is rather prickly, you will want to shred the leaves very fine before adding a mild French dressing. Other leaves may be put in the blender with other fresh greens to add minerals and vitamins to the green drink you make. Or pour boiling water over some young leaves to steep for 12 minutes for an herb tea. Rolled, tied comfrey leaves added to (and later removed from) stews or soups can add nutrients. If you wish, you can freeze or can young comfrey leaves for use later. If canned, process for 30 minutes in sterilized jars. For freezing, cook first. You can also dry young comfrey leaves in the sun or in an oven at 125°. Dried, pulverized comfrey makes a fine addition to the herbs you put in your salads the following winter.

dill

anethum graveolens

Will grow in any average open area with a little compost dug in along the row, as long as the soil is not acid. A packet of seed will sow 10 feet and give you plenty of delicate leaves to have fresh in salads, and some seeds for pickles later. The crop of seeds matures in 70 days. Keep the plants watered in drought time to discourage pests, and spray with a garlic-onion blender mixture, if you see that red spider mites are getting on them. If your climate is not too severe, plant some seeds in the fall, so they will come up early the following spring. Some people boil dill instead of caraway with cabbage, turnips, and cauliflower. A good trap plant for green tomato worms.

fennel

foeniculum

A pleasant herb to have for salads—if you like anise as a taste. The leaves and tender stalks can also be eaten boiled, with or without a cream sauce, though the bulbs are the part usually cooked. If left alone, wild fennel (*Foeniculum vulgare*) will mature in 90 days and grow to 5 or 6 feet, come to seed, scatter seed, and grow up just about everywhere the following year. Therefore, cut back the stalks two or three times, but toward the end of summer let some self-seed. Black fennel is a very handsome plant and tastes just about the same. Try fennel tea, made by pouring a cup of boiling water on a teaspoon of bruised fennel fruit, or seed. Fennel is one of the Nine Sacred Herbs (along with mugwort, plantain, watercress, camomile, nettle, chervil and crab apple).

Another kind, Florence fennel (*Foeniculum dulce*), has whitish stalks like celery which you harvest as a bulb and braise. Get the variety called Mammoth to sow in late May or June, spacing 6 to 8 inches apart. In Italy, where the bulb is usually braised, it is called *Finocchio* and has often been used to stuff keyholes to keep out ghosts. Otherwise it has been cooked with fish, used in fish sauce, its seeds added to sausages and soups, rolls and breads, cooked with beet greens, and ground to a medicinal powder for comfort to the stomach. Most seed houses carry it.

feverfew

chrysanthemum parthenium

A prolific tangy herb, very useful as a repellent. It has pleasant little white flowers that look like small daisies. Herb and flower houses carry it; or dig it up along the roadside, where you often see it growing in the gravel.

garlic

allium sativum

Is just about the most useful plant an organic gardener grows both for its excellent taste and because of its potent and efficacious antibiotic and repellent qualities. It has an incredibly ancient history of use by the peoples of Europe from Siberia to Sicily. In some polite circles it has been thought vulgar, partly because the peasants in various cultures regarded it as an aphrodisiac and its smell on the breath was always the giveaway. As early as the sixteenth century it was used as a mole repellent ("to make them leap out of the ground"). For modern pest control plant cloves of garlic around your peach, apple, and pear trees, and here and there among your vegetables. Grow a row so you'll have plenty for your own use in the kitchen and enough to make sprays and to transplant more of the cloves whenever needed. Garlic does best in well-prepared, enriched soil. Divide bulbs into single cloves and bury them two inches, though as a repellent you may decide to put some nearer the surface. Weed the garlic plants well, dust the rows with soot, wood ashes, or compost, and be sure you plant in full sun. The 1971 report from the University of California at Davis said that the antibiotic in ten parts per million could be effective in controlling pests in your tomatoes or corn.

The virtue of garlic is in its sulfide of allyl, an oil present in all members of the onion family. It is so good a protection against unwanted bacteria that in wartime it has been used as a disinfectant (on a swab of sterile sphagnum moss). It is very health-giving and is liked by cattle, dogs, fowl, gorillas, and wild animals who seek out the colonies of wild garlic. Perhaps you have tasted spring milk that is redolent of the garlic the cows ate.

horehound

marrubium vulgare

The leaves of white or common horehound are used for flavoring candy and making teas. You can buy the seeds from Burpee's or Nichols (who also sell plants) and save a few of the 18-inch plants for a perennial. Horehound likes poor, dry soil and does well if sown in spring or increased by root cuttings.

Black horehound (*Ballota nigra*) is a good fly repellent and efficacious, so they used to say, "if you are poisoned by your stepmother." Whereas white horehound has a pleasant odor, black horehound has a foul smell.

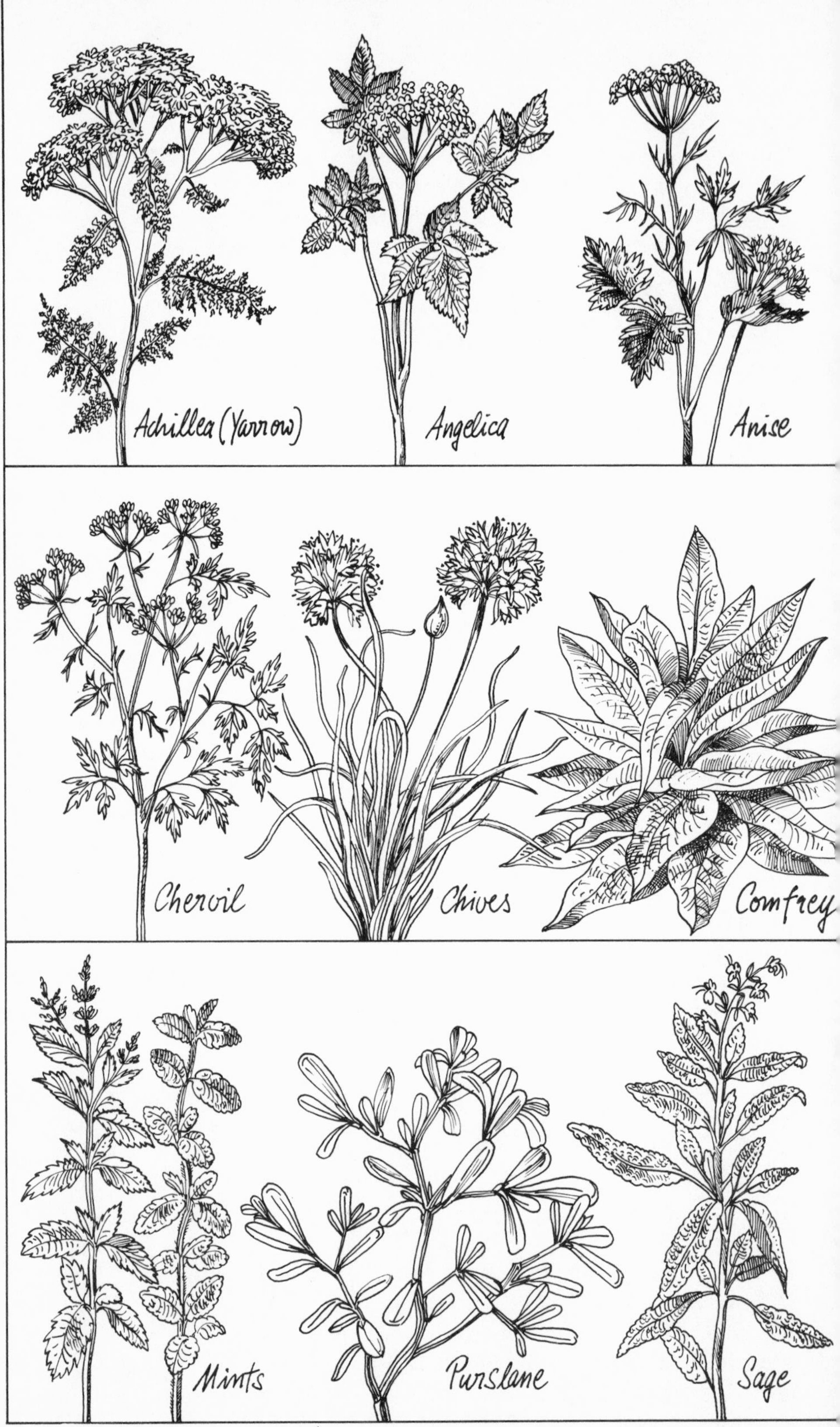

Unusual Herbs and Weeds

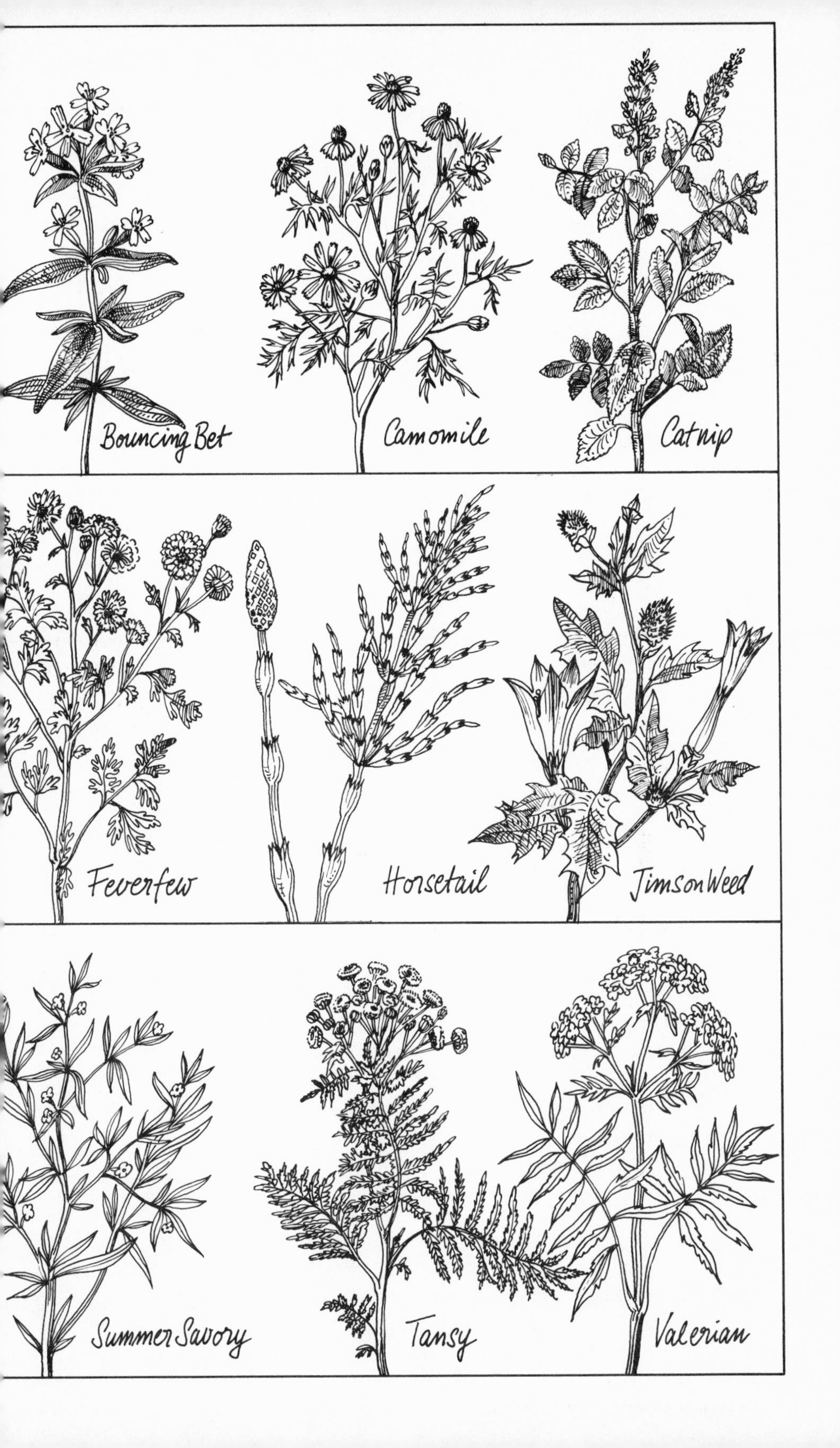
Bouncing Bet
Camomile
Catnip
Feverfew
Horsetail
Jimson Weed
Summer Savory
Tansy
Valerian

horseradish

armoracia rusticana

If you can get hold of a few pieces of root, you can start right out on your horseradish plantation. Put it in rich soil with manure dug down 18 inches into the soil, water it well, and you'll have all you need in short order. Until the Germans invented horseradish sauce in the eighteenth century, to eat with Friday fish, it was used for medicinal purposes only. The best way to have a year-round supply is to dig some roots to winter over in a box of damp sand. This herb is a violent spreader. Watch it carefully.

horsetail

equisetum hyemale

The scouring reed, known throughout Europe as the best means for cleaning pewter and other metal pots. This is because it takes up into its own system the silica which most other plants reject. It comes from a very early point on the evolutionary scale, and can still be seen growing along the gravelly rims of streams, marshes, and roadsides. There are two forms: the many-branched one that looks like a horse's tail and the straight one with no branches at all. I once saw horsetail offered in one nursery catalog, but if you can't find some plants yourself in the gravel beside a road, I think an accommodating all-purpose nursery would help you out if you want to grow your own Chore-boy.

hyssop

hyssopus officinalis

This rather bitter-tasting herb with square stems is added to pickles, poultry stuffing, and meat pies, especially combined with parsley and sage. Some believe that it "slayeth worms in man" and will "heal all manner of evils of the mouth." It is used in flavoring the liqueur Chartreuse. Hyssop is attractive in the garden, for it is almost evergreen and has a growth habit like a shrub. In the old days it was used to make mazes, and propagated either by cuttings or by seed. It prefers a dry, light, warm soil, with plants spaced a foot apart. This is a very strong, pungent herb. Try a little piece in stew. For catarrh, combine it with white horehound, and infuse it with two cups of boiling water to the ounce. Add honey. Bees love this plant.

jewelweed

wild touch-me-not
impatiens capensis

A good wild vegetable if eaten young, buttered and peppered, or creamed on toast. It is a very soft and succulent plant, growing in moist places. The Indians used it to make a juice to rub on itchy spots. It contains a fungicide, which was isolated at the University of Vermont by Thomas Sproston, Jr. Euell Gibbons has high praise for the juice of this plant as a prevention for poison ivy rash. Use it liquid, or make into ice cubes and keep in the freezer to apply when you have been or will be exposed to poison ivy.

jewelweed . . .

marigold

tagetes minuta
or
calendula officinalis

Either fresh or dried, the flowers of marigold have been used since olden days for soups, drinks, and custards. To apothecaries, the virtues of this plant included its powers against the pestilence, its ability to draw evil humors out of the head, its efficacy against intestinal troubles, and its curative properties when rubbed on scabs. Down into the eighteenth century marigolds were used for lotions and ointments and for binding up skin afflictions. Today the marigold, especially the small-flowered Mexican variety, *Tagetes minuta* (available from Thompson and Morgan in England), is highly valued by the organic gardener as a deterrent of nematodes and other pests in the garden. I plant marigolds in among the vegetables, around the tomatoes, and on the outer edges of the plot, along with nasturtiums.

If you can propagate a white marigold, you can win a huge prize. (Let Burpee know.) The name, which sounds like the word *calendar,* reflects the old belief that these flowers can be found blooming somewhere in every month of the year. Other uses of marigolds have included yellow hair dye and medicine for cuts, inflammation, heart trouble, and warts. It is a very hardy, pestless, unfussy plant, and should be grown plentifully in every organic garden. It can be planted outdoors in the ground, and slightly thinned when it gets to be three inches high. It can also be started indoors any time in March or April and be ready to set out after all danger of frost. It is sensitive to frost and in the fall will turn dark green and brown at the first drop below 32 degrees. In mild climates it will self-sow and grow up all over. Save seeds yourself. If you live in a colder climate, leave one plant of the tall variety in a big pot in the house. Fertilize it well, and it will turn into a huge house plant. On a pinholder in an arrangement, marigolds will root and last for weeks.

marjoram

origanum

This plant has been favored by man for many uses for many centuries. In general, it is a perennial and will winter over in some forms in the ground if it gets a good start. It favors *marjoram . . .* a limy location and will grow to 18 inches when the soil is just right. Mine always grows much smaller, both the mild annual kind in the garden, sweet marjoram (*Origanum marjorana*), and the pungent kind with dry purple flowers that grows in the fields, the perennial wild marjoram (*Origanum vulgare*). The wild kind is not good in salads and stews, but it makes an acceptable tea.

For sweet marjoram, the delicious native of Portugal, harvest the leaves as needed throughout the summer, and then the whole stalk just before flowering to make dried herbs. It is excellent in salads, omelets, soups, and stews.

milkweed

asclepias syriaca

For several years I had to go out to look for milkweed, but now it has arrived in our garden, and I am glad of it. From the first early shoots right through the time of flowering there is something good to eat on the milkweed plant. The tips come back if you cut them off, and for several months they make tender tidbits. The flowers, though somewhat flabby and sometimes bitter, do make good tempura (see p. 139). They are very pretty in a platter of mixed tempura herbs and weeds. The pods can be cooked, and if you let some go to seed, the silky wings will fly all around and settle down in the grass, in the garden, and along the edge of the road to bring you new plants the next year.

In order to avoid the bitter taste, which might annoy you, cover the part of the plant you harvest with boiling water, boil for two or three minutes, then recook in a second water. I find that milkweed cooked in the pressure cooker, then drained, gets nearly as good results. I always cook only the top three inches, and with a good buttering or oiling with soy oil and lightly salted, they taste to me like a cross between asparagus and spinach—though if the bitter principle is not drained out, they can taste more like cooked dandelion. The young pods, picked while still very tender, are a special gourmet delight. They are easy to freeze for winter use.

mint

mentha

You can grow spearmint (*M. spicata*), peppermint (*M. piperita*), apple mint (*M. rotundifolia*), and others. Though it likes moisture, mint will do well in any good soil, and it may spread through the garden into the fields and all sorts of corners.

It is very helpful to keep a section of your mint beds well weeded and loosened so that you can dig some up on short notice to move to a trouble spot in the vegetable garden as a pest repellent. An infestation of aphids, for instance, can be curbed sometimes by introducing mint along the row. After hosing the infected plants, set in the slips of mint, and leave them there as long as convenient. Mint is rugged and does not mind being hauled around. Mint sauce, jellies, juleps, and garnishes are all easy to make.

nettle, stinging

urtica dioica

For a year I tried to make up my mind to cook and eat some stinging nettles, for I thought if Euell Gibbons did it, why couldn't I? Yet when spring weather came, I let the crop get large and go past its tender stage without making an effort to put on my gloves and harvest it. Maybe some other year. Who knows? Sir Albert Howard recognized that the deep probing roots of the nettle make it valuable as a soil improver, and also a very nutritious herb to throw on the compost heap or to feed the hens. It is full of nutrients for the brave soul who will pick, cook, and eat it. In Scotland, until recently, it was used for fiber, and many a household had nettle sheets and nettle napkins. Obviously the stingers have been removed.

To cook nettles, have the water boiling, and throw in the tender tops of the young plants. Do not use much water. Cover and cook gently for twenty minutes. The leaves can also be blended in water for a good spray—either alone or mixed with garlic, marigold, or mint. You never see pests on nettles. Therefore, follow the hint from nature, and spread bits around on those plants that you want to protect.

Some old remedies include whipping arthritic sore places with sprays of nettle. Nettles also provide a kind of rennet to curdle milk.

parsley

petroselinum crispum

A well-known herb, and a good thing, too. Its high vitamin C content, especially in the stems, can keep the scurvy away for anyone who will bother to have a row of pots of it on the windowsill all winter. It also has lots of iron, so wards off anemia too. It is easy to grow in the garden, easy to adapt to flower-pot life, easy to plant outside again in the spring. You can grow the curly-leaved variety, which is often used as a garnish, or the darker flat-leaved variety, often called Italian parsley, which is more aromatic. Since it is a biennial, it will bolt the second year, but there are still plenty of side leaves, and I use the blossoms for tempura. Its oil, called *parsley . . .* apiole, comes from the seed. Its roots are dried for tea, and so are its leaves.

Do not play around with wild parsley. It may turn out to be poison hemlock or what the English call fool's parsley. The look may be deceivingly the same, but the root and the smell are quite different. If you feel tempted, bruise the leaves and smell the juice. Once you've smelled parsley, you'll know true parsley again. Only take what smells that good.

purslane — *portulaca oleracea*

A good weed. It will be all over your garden, growing low to the ground, with round smooth leaves, so don't bother to plant it. The young shoots are good in salad, almost as full of iron as parsley; the older ones can be cooked like spinach. Used in old times as a mouthwash.

rosemary — *rosmarinus officinalis*

A good bee plant that will grow rampant and be a perpetual emblem of remembrance, as Shakespeare said. It is used in incense, for sprinkling on buttered crackers for hors d'oeuvres, to make lutes, to ward off the evil eye, and for Christmas decoration along with holly and ivy. In a warm protected place in England rosemary can reach tremendous heights and live for twenty years. It is a half-hardy perennial in this country, growing to 3 feet in eighty-five days. It repels pests, and I use it in salads, stews, scrambled eggs, and bouquets garnis.

sage — *salvia officinalis*

There are various kinds, including golden, pineapple, and purple-leaved, but the perennial called broad-leaved is the most satisfactory for the home gardener. Start a row, thinned to six inches apart, from a packet of seed, and then move the plants around to convenient spots in your garden where they can remain as pest repellents. They will grow about twelve inches high.

For chicken stuffing, says a 1780 cookbook: "Take parsley and sage without any other herbs. Take garlic and grapes and stop the chickens full and seethe them . . . and mess them forth." How does that sound? Sage makes a soothing tea, especially if milk is added. It is excellent in sausages, and can be eaten raw as a welcome refresher.

The kind of sage called Clary (*Salvia sclarea*) was a preserver of the clear eye, it was thought in old times, and was taken in a way "fried with eggs in manner of a Tansie" (see below). This evidently means a tempura, for other old recipes mention dipping sage leaves in a "batter made of the yolkes of egges, flour, and a little milke, and then fryed till crispe."

sage . . .

summer savory

satureia hortensis

Matures in sixty days, for a dried herb, but can be snipped earlier to make *fines herbes* with basil, chervil, tarragon, and rosemary. It is an annual raised from seed, and should be thinned to six inches apart and well watered. It goes especially well with English broad beans and soybeans. A good way to preserve it is in an herb butter (3 tablespoons to ½ pound of salted butter).

tansy

tanacetum vulgare

The young leaves, mixed with eggs, were known as a "tansy"; this was often eaten as a Lenten dish, believed to be good for cleansing the "bad humors." For centuries a tansy tea has been used for stomach cramps. The great value for the organic gardener is to plant it in clumps around the garden to ward off flying insects, especially those coming in to lay eggs. It has a fine, strong, very pungent odor, and is a tall, handsome, feathery-leaved plant. Near the door, it repels ants.

tarragon

artemisia dracunculus

A delight to the gardener and cook. Its cousin, *Artemisia redowski,* called Russian tarragon, is a bigger, handsomer plant with a much milder taste and bigger, juicier leaves. If you buy seed, it is this Russian kind you get. The so-called true tarragon, which comes from places like Tartary and Chinese Mongolia, produces no viable seeds and has to be propagated by root cuttings. These cuttings are tender and fussy and should be ordered in the spring or early summer so they will establish themselves well before the winter. By ordering early, you also give yourself a chance to reorder if the first planting fails. True tarragon grows about two feet tall, and spreads about two feet. Russian tarragon grows three or four feet and spreads six feet.

Either makes a good tarragon vinegar (soak leaves in cider vinegar one to two weeks), though the Russian tarragon vinegar is a good deal milder. The mild fresh leaves are good in salad, sandwiches, and cooked with chicken, but the stronger true tarragon is much better for drying.

thyme

thymus vulgaris

Often used for a potherb or combined with honey as a cough syrup. Very tangy and nice in stews, meat loaf, and most soups. The thymol in it is an active ingredient in cough syrup. Bees love thyme, and it makes a very interesting honey. Strong enough to repel many pests. There are many varieties.

woodruff, sweet

asperula odorata

Grow a little to put in your May wine, and to drive off moths when you strew it around the closet. It is cousin to an escape called lady's bedstraw (you may have it as a troublesome weed), which was used in old times as a cheese rennet and to put in a mattress to make a lady fertile. The Greeks used woodruff, steeped in oil, to anoint the tired-out traveler who came to study herbs with the wise men in Greece. (Herb study, gardening, alchemy, even Christianity, were inextricably interwoven in people's minds throughout the Middle Ages, and it was believed that only the pure in heart could succeed in any of them. Such ideas put a damper on the exploitation of gardening for commercial gain and developed the meditative, devoted gardening expert who studied the virtues of plants and the interrelations of plants and man. He often discovered values and secrets now well explained by modern physiologists.)

wormwood

See *Artemisia,* p. 118.

yarrow

achillea

A white kind grows wild, and you can order varieties with yellow, white, or red flowers, called achillea or milfoil, from the nursery. Yarrow is quite tangy and beneficial as a protection plant. Bring some into your garden and watch the pests fly the other way.

kale

brassica oleracea acephala

Very good cabbagy greens for late fall and early winter. Also they make a lovely, crisp relish when the leaves are picked very young, to put as a decoration on an hors d'oeuvre tray. Vates, dwarf blue curled kale, is the best to grow, and it will mature in 55 days, so you can keep on sowing it well into July. Plant after peas, for instance. A slightly more delicate kind, called Siberian, matures in 65 days. In zones south

Fruits and Vegetables high in Vitamin C

kale . . . of New Jersey, you can keep kale in the ground all winter, so you can sow it until September. Not a bit fussy about soil, but a heavy feeder, especially when young and growing fast. For best results use both manure and lime when preparing the ground. Do not plant near cabbage.

The pests that might attack kale are flea beetles, aphids, and the female Herculean beetles which come to lay eggs. Good provision of nutrients, good mulch, and herbs planted nearby will help kale plants. Recommended are mint, thyme, catnip, hyssop, and rosemary. In the rare case of a terrible infestation, try rotenone.

Start harvesting from spring plantings early in June. Eat raw or cooked. There is a good supply of riboflavin and of vitamin A in kale, with more in the green parts of the leaf than in the midribs and stems. To retain kale's good supply of vitamin C, wrap the vegetable on picking and store in the dark in a crisper drawer. If you boil it, use very little water to keep in the 90 percent of the vitamin C you can save by so doing. Macrobiotic, yoga, and USDA recommendations all stress that cabbagy and other green-leaf vegetables should never be cooked beyond the point where they are still crisp. Sautéing in oil first is recommended for all these succulent vegetables.

kohlrabi

brassica oleracea gongylodes

This member of the cabbage family needs more calcium for its growth than any other, and dolomitic or calcific limestone should be applied to the area to be planted at the rate of 75 pounds per 1,000 square feet a week or so before sowing the seeds. In the right climate you can get an early crop by planting in March or April and another before frost by planting about the first of August. When the bulbs are well developed, dig them and eat. They grow above ground, so you can see when they are ready.

A good variety is Early Purple Vienna, flushed with pinky purple. It matures in about 55 to 60 days. Another variety, Early Purple, takes longer and grows larger. If you harvest some when the bulbs reach about 2½ to 3 inches in diameter, you will find them quite different from the vegetables you let grow to big size. A packet of seeds will sow 40 feet, and the plants should be thinned to 6 or 8 inches apart. They are good sliced very thin and eaten raw. Japanese beetles can be a problem, but use the standard tricks to cope with them. Introduce into the ground, especially in the lawn, at three-foot intervals, a teaspoon of *Bacillus popillae* or milky spore disease (see Appendix), which is fatal to the larvae of these beetles, but to no other creature. Ladybugs and praying mantises also might help, for they eat larvae. Remember that moles and skunks eat the larvae too.

leeks

allium porrum

This delicious mild member of the onion family is complicated to grow, but absolutely unmatchable for soup and braised hot or cold and eaten as the Greeks do. Anyone who likes onions also likes creamed leeks on toast—though unless very tender they can seem stringy. Buy plants or start seeds yourself outdoors. Plant seeds thinly, half an inch deep, after the frost is out of the ground, or indoors in a flat or in the cold frame. When they are 6 to 8 inches high, dig them all up, cut back about half the top growth, and replant at a distance of about 6 to 9 inches apart in 6-inch trenches, which you fill up as the leeks grow. Plants bought from a nursery are handled the same way. Leeks are not fussy about their soil, but if you give them a good dressing of compost, keep them moist, and fill the trench gradually but persistently, they will grow big and sweet. To keep the stem white all the way up to where the leaves branch, keep on banking with earth after the trench is full or use a paper collar which you gradually push up as the plant grows. This seems like a lot of work, but the only leeks worth growing are those with good, tender, thick, white stems, and this is the way to get that superior quality.

Leeks may be balled up and kept in a box over the winter. I have even balled them and kept them in a protected place outdoors over the winter and had them there ready to start using when the snow melted. I have also grown leeks the lazy-man's way, by just planting the seed in the garden and letting them go. They do not amount to much that way, I can assure you, but even so they make marvelous soup.

The proportions for soup are two large leeks or eight small underdeveloped ones, one medium-sized potato, and three small sprigs of marjoram. Chop all fine and cook in a very little water until the leeks are tender. Then blend this mixture, adding a little chicken stock if needed. Put the rest of 2 cups of chicken stock in a glass double boiler, add 2 cups of milk and ½ cup of cream, salt, white pepper, and a pat of butter. Simmer until the flavors are mellowed, chill, and serve with a sprinkling of chives on top. You may think *leeks . . .* you have a vichyssoise, but actually it is a soup with the most elusive, mild flavor imaginable.

For creamed leeks, sauté in butter, simmer, and add a cream sauce made from unbleached white flour and cream. Serve on toast.

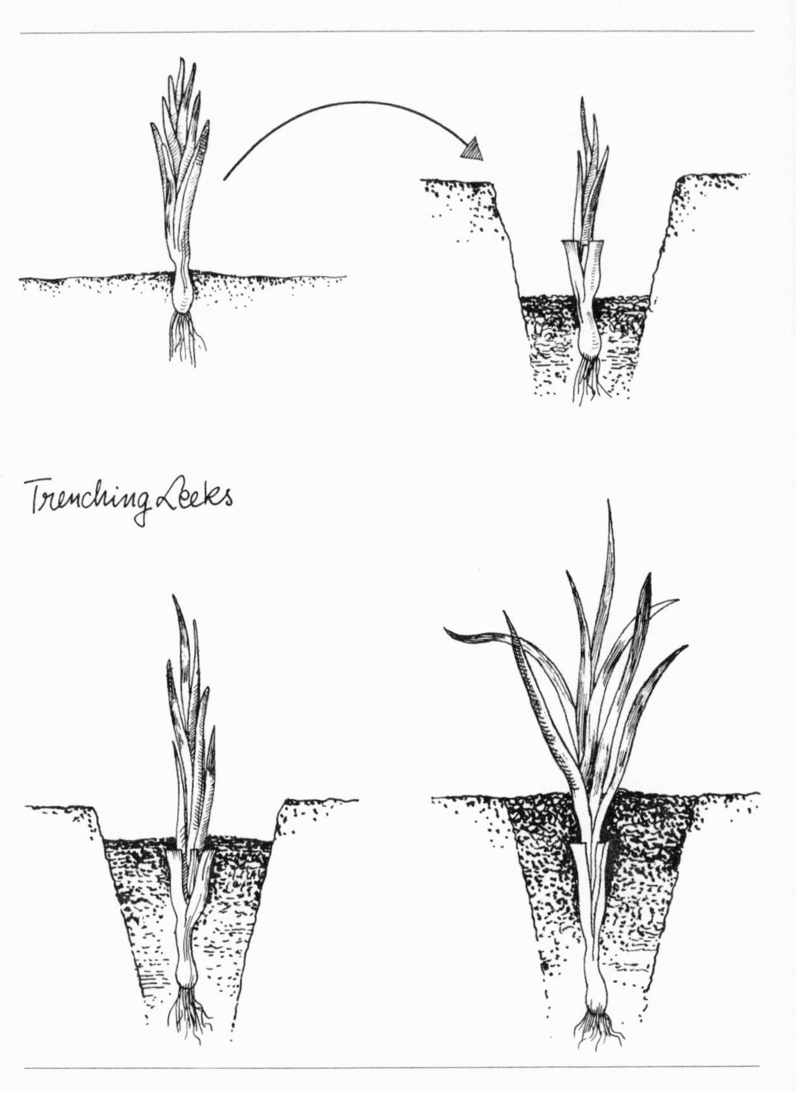

lettuce

lactuca sativa

You can grow twenty kinds, and you'd probably find them all good. Since a packet of seeds grows too many plants at once, the only way to avoid overplanting and crowding is to time them, plant thinly (preferably mixed with radish seeds), weed, transplant, and then plant a few more as long as the weather is somewhat cool. Lettuce doesn't germinate and grow well in hot weather. In the old days, herbals presented four kinds: garden, curled, the cabbagy kind, and Lumbard lettuce. Often used for a boiled green, it was believed to be more digestible that way, though we know today from nutritionists that the minerals and vitamin C in fresh, uncooked lettuces are preferable.

Lumbard lettuce is loose-leaved; the cabbagy kind is romaine or cos; and the curled is escarole. Early settlers believed in starting a meal with lettuce, but they also recommended it for nibbling after dinner, too, to "stay the vapours of wine from rising up into the head." They also knew what the great English herbalist Nicholas Culpeper said: that the Moon owns lettuce, and that it in all ways "cooleth and refresheth," and has a very good effect "upon the morals." It was served with unblanched endive, chicory, and purslane, in a sauce of vinegar, oil, orange juice or lemon juice, salt, pepper, and sugar—which certainly still sounds familiar as a dressing.

A partially shady place can be used for head lettuce as long as the soil is loose and rich. Leaf lettuce does well in full sun. Since it is very shallow-rooted, you need only rake in an inch or so of compost, and add some along the rows when the lettuce is up and properly thinned. This is one plant you can safely water in dry weather without injury to the roots. Do it early in the day, so the sun will dry the plants well before evening. In fact, lettuce roots are so near the surface that it is the first plant to wilt during a drought. This root habit also makes it advisable to do any transplanting at the end of the day, after the sun has gone down. The rows need only be 12 inches apart, but the plants should be spaced 8 to 12 inches apart for best heading. If you plant small areas of lettuces all together according to the French intensive method without rows, keep the plants dry and the slugs all picked off.

Experiment with different varieties for several years to find out what suits you best. I like oak leaf lettuce very much, both as a fresh young lettuce and as a mature plant later in the summer. The young leaves wilt rapidly, so use it as soon as you pick it. Slightly more sturdy are the Black-Seeded Simpson varieties and a 40-day favorite of mine, Salad Bowl. This is slow to bolt, so you don't have to race to eat it up before it develops a tough stem. A pretty variation is the reddish Ruby lettuce, which is very tender, too. We do not grow iceberg lettuce, but I believe that the kind called Great Lakes is reliable. A butterhead lettuce called Matchless *lettuce . . .* is very good. We grow Bibb, Summer Bibb, and Buttercrunch, and find them all excellent.

A big, crisp lettuce is the 76-day cos or romaine lettuce, with tall, dark green leaves growing upright and forming 10-inch heads with white hearts. It is more crisp than the delicate oak leaf lettuce, but it makes a good contrast in a mixed salad.

The most long-lasting in our garden is the curly endive, or escarole, which has a 95-day maturing period. We pick and eat it long after the first frost in the fall. A less curly kind, Florida Deep-Heart, is also good until fairly late in the season. Its leaves are quite tough and withstand adverse weather. Sow these big plants in rows 18 inches apart and thin to 18 inches, too. Lettuce of several kinds should also be started in your hot bed or cold frame, beginning your plantings in March or April and keeping them up until the end of May. You can start them over a heat coil, move them to a cold frame, and eventually outdoors to the garden. Often the transplanting from a cold frame is easier than thinning and transplanting in the garden, especially if you use such modern aids as Ferto-pellets and Ferto-pots. Always thoroughly bury such containers so they will stay moist. Otherwise they block and kill small roots when they dry out.

mustard greens

brassica juncea

Relative of cabbage and all those *caulis* plants. The mustards grown for seeds are other varieties. I don't know what keeps northerners from growing mustard greens—either indoors or out, the young, tender leaves make a fine addition to a salad, and the cooked greens are a good variation from spinach. I grow a flat in the kitchen, and it is very handy to have there. Sow the seed a little at a time, to keep the crop coming. Outdoors, this can be done until the weather gets quite hot. Recommended varieties are Southern Curled Giant or Fordhook Fancy. If you want one that tastes quite a lot like spinach, try Tendergreen, which is more heat- and drought-resistant than the others and grows fast. Let a few plants go to seed in order to harvest the seed for future crops, but be sure to pick them before the pods open, and in dry weather.

Homemade Mustard: In a mortar pound 1 tablespoon chopped parsley, 1 tablespoon chopped tarragon, 1 teaspoon chopped chervil, and ½ cup mustard seed. (Of the store-bought mustard seed, the black is more pungent than the white or brown.) The grinding can also be done in the blender, but the flavor may seem more dull. Put this through a sieve and save the powder that comes through. Then slowly add 1 tablespoon of oil, preferably corn oil, and 1 tablespoon vinegar. Stir and keep covered. Instead of vinegar you can use a tart juice such as currant juice. Use up homemade mustard quickly, and then make a new batch, for it does not keep well.

marjoram

See *Herbs,* p. 167.

nasturtium

tropaeolum

There are several varieties that will do well as companion plants in your vegetable garden. Be sure to include some yellow-flowering ones, as repellents for aphids, which have a reaction to yellow. Nasturtiums can be sown outdoors in May, after the ground is warm, or started inside in individual Ferto-pots. They do not like cold, damp ground, but they blossom very well on poor soil, so plant them along the edges of the garden where you have not put very much enrichment and compost. Where the soil is rich, they go to green leaves, which, though pest-repellent, too, are still not so serviceable for the gardener as the plant that gives blossoms and seeds for eating.

For varieties try Golden Globe, Golden Gleam, Primrose Gem, and collections such as Mixed Double Dwarf, Mixed Dwarf Single, Tall Climbing Fordhook if you have trellises, and Dwarf Jewel Nasturtiums, which are especially prolific in bloom. Plant them between the vegetables for pest repellents in places where you don't plant marigolds and the other useful flowers.

okra

hibiscus esculentus

Familiar in southern and western gardens. It is possible to grow okra in the Northeast, too, especially if planted so that it grows in the heat. Plant the seeds an inch deep in good, friable soil, well drained. If the soil gets cold and wet, the seeds will rot. In a spell of bad weather, change any plans you have about planting it outdoors, and start it indoors or in the cold frame or hot bed instead. Be sure it gets plenty of nitrogenous nutrient such as rotted manure and compost or leaf mold, preferably applied a month before planting. Okra will grow big and bushy, so plant in hills three feet apart, keeping the seeds several inches apart. In nonorganic gardening books, the distance recommended is 12 inches *okra . . .* apart, but it is obvious that with a nutrient program like that practiced by the organic gardener, it is only sensible to leave room for big growth, especially of woody plants like okra.

When the pods begin to mature, and get to be a few inches long, start harvesting. Remember that they should be only half-grown when picked. Use them as many times as you can for the splendid Louisiana gumbos—shrimp or chicken. Freeze if necessary. And eat some as a boiled vegetable dipped in hollandaise sauce if you don't mind the slipperiness. A fine source of vitamin A, calcium, and phosphorus.

onions

allium cepa

The globe onion can be grown from seeds, from purchased plants, or from sets (the small bulbs that can be bought from the seedsman). With good soil they all will mature for home use during a summer season, but at different times and with somewhat different qualities.

If your topsoil is deep, loose, humusy, and a bit sandy, onions will do well. You can start with sets, of the Ebenezer type, planting them where you want the final onions to be. If you use a row, put them about three or four inches apart unless you plan to eat some very young, and cover with one-fourth inch of soil. One pound will plant a 50-foot row. In five weeks you can pull them as scallions. Later in the summer, when the tops brown or fall down, harvest the crop. The plants to send away for are Ebenezer or sweet Spanish. Plant them about two weeks after you put in the sets. Ebenezer onions are especially good for winter storage, so don't dig too many for summer use. Those you grow from seed will come along last, for it takes 130 days, more or less, for them really to mature. Harvest onions after the tops die down, and be sure to dry them as much as possible in the warm sun.

The great convenience of growing onions is that there they are in the garden, full of flavor, and ready to pull in some form or other all summer long. If you plant so-called multiplier onions, you can have them even in early spring. The ones grown from seeds can be pulled for a substitute for chives at any time after they emerge from the earth. You don't have to grow all three versions, and many people settle for only one.

The hazard for young onions grown from seeds is weeds. Therefore it is sensible to grow them in a part of the garden fairly free from weeds the year before. You can plan to have them follow carrots, beans, or corn. Keep them weeded, and do not let any caking of the ground occur. The little roots are very easy to injure, so weed or cultivate with the greatest care.

The pest to watch out for is the onion thrip, a tiny fly that lays its eggs at the axil of the leaf and whose nymphs suck the plant sap. A soapy spray or rotenone (4 percent) helps to fight this pest. Keep an eye out also for the onion maggot, the larva of a gray fly like a house fly, which works down into the onion bulb. An oily-soapy spray on the ground will help against this pest. The editors of *The Organic Way to Plant Protection* recommend that you spread your onion plants all around the garden to fight this maggot, instead of having them in a row. Good advice. By so doing, you also offer onion protection to plants whose pests are repelled by this aromatic species. We put onions all around the garden to repel rabbits and woodchucks. We also use pungent flowers, wood ashes, and sometimes (after many experiments) blood meal.

The one kind of onion growing we have settled for after many experiments is from seeds. They are cheaper, and very little trouble, and we cut down on bringing in diseases on purchased plants or sets.

parsley

See *Herbs*, p. 169.

parsnip

pastinaca sativa

This vegetable is called a cool-weather, long-season plant, because it grows well in northern states and is able to live over the winter so you can start digging it the minute the snow melts off. Parsnips are easy to grow and very nutritious. For years I thought I did not like them, but recently we have started growing parsnips for harvest mostly in the following March, and now it seems like a rare treat to have braised parsnips fresh from the ground for dinner in the spring, or slivers of raw parsnip with a sandwich for a springtime lunch.

To get straight, unforked roots the ground should be loamy and loose, dug eight inches deep and composted. Plant seeds a quarter to half an inch deep at the same time you plant radishes. In fact, mix some radish seeds in with the parsnip seeds, so you will know where the row is coming, for it takes parsnip seed a long time to germinate and come up. After the plants are well established, thin to six inches apart and cultivate and weed them until they grow up so their leaves touch. After this they are no trouble at all. We

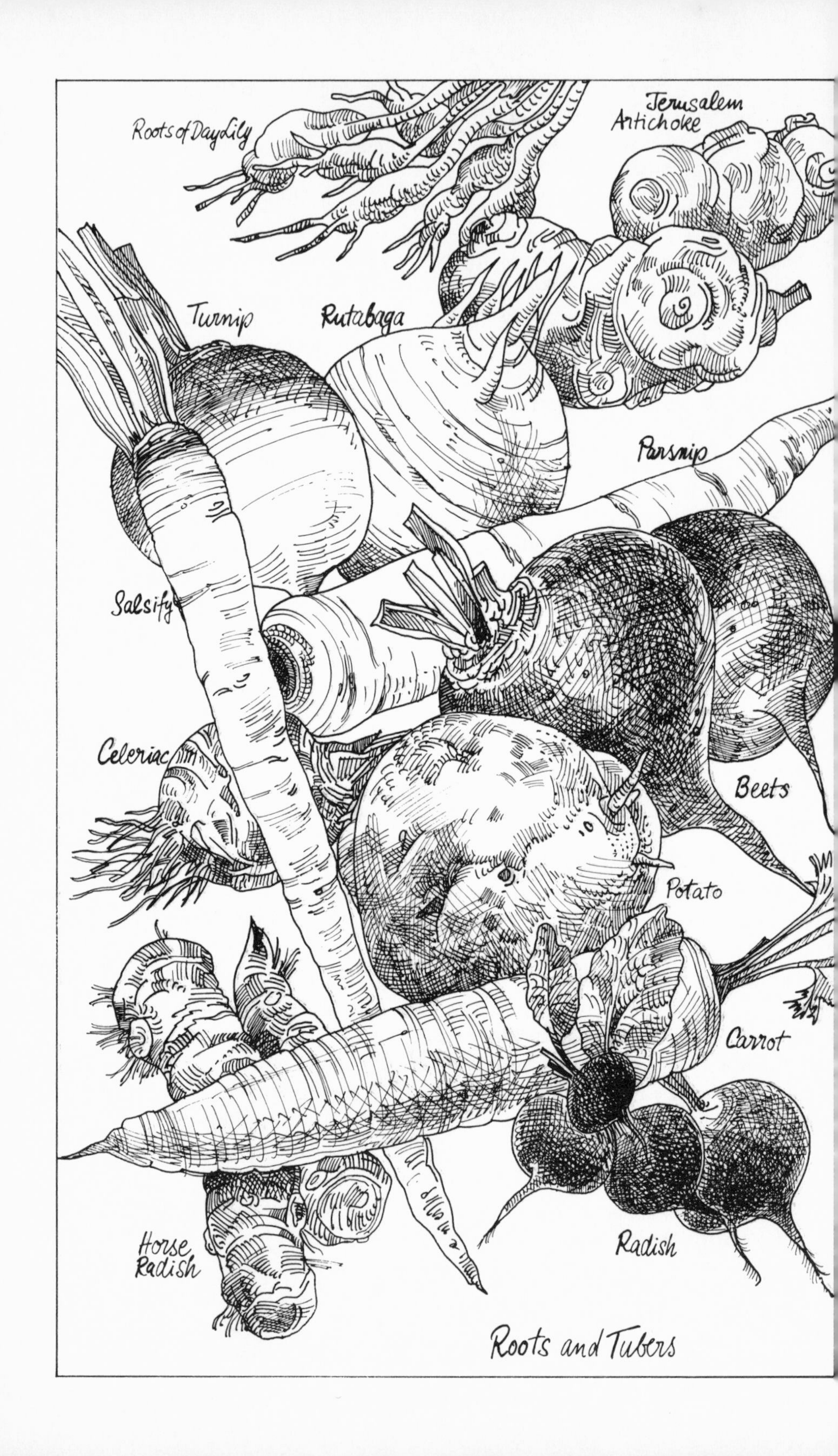

Roots of Daylily
Jerusalem Artichoke
Turnip
Rutabaga
Parsnip
Salsify
Celeriac
Beets
Potato
Carrot
Horse Radish
Radish
Roots and Tubers

like them to be about two or two and a half inches in diameter at the top, so we do not fertilize them very much. All American and Hollow Crown are the varieties that are best. These take 95 to 105 days to mature.

Be sure to try some parsnips in the fall. Parboil them and then finish the cooking in melted butter or oil. If cooked very slowly and for quite a long time, they become nutty and browned. There is enough sugar in them to give them a carameled quality.

When you leave parsnips in the ground over the winter, dig them as soon as the frost melts and eat right away. Leaving them in the ground too long makes them turn to new plants and their tops grow. If you let them sit indoors, they dry up and get corky and tasteless even in the crisper compartment. Everything else I can think of about parsnips is good. In a cup of parsnips there are five more milligrams of vitamin C than the daily requirement, and plenty of minerals.

For companionate planting, grow parsnips near peas and beans.

peas

pisum sativum

An old gardening book I have starts its entry on this vegetable by saying: "Green peas are the gardener's pride and joy." They may well be, especially on the first day of the season when the gardener brings to the table the young, delicate, sweet, tender peas either raw in salad or just barely cooked, and within ten minutes of picking them in the garden. Other peas in comparison taste weeks old, overcooked by half an hour, tough, tasteless, and worn out. Besides, very fresh peas are highly nutritious.

Peas like coolness, so get them into a well-prepared, moist soil as early as possible. If you use wires or brush sticks to support them, put them up early to be ready for the seedlings when they emerge. Make a trench along both sides and plant your peas in alternating spots on either side of the support. Fill these up with six inches of compost which you will push aside in order to plant the seeds near the bottom of the trench. Put in extra nitrogenous matter, such as cottonseed meal, fish emulsion, fish meal, or something like peanut, sugar, or cocoa wastes. Also add some lime. Get the soil ready a week before planting if you can. But never tramp down soil that is wet and heavy; it could ruin it for the whole season.

Varieties of climbing peas to use are Little Marvel,

which matures in 63 days, and Alaska, which comes even earlier, at 55 days. For a very tall pea, you can try Alderman, which comes later, in 74 days. All are heavy bearers. For bush peas, try Burpeeana Early, at 63 days, and for a late pea, Fordhook Wonder Wando, or Lincoln, at 67 days. Wando is the pea best suited to warmer climates or late planting. But keep the soil moist so it will germinate.

Be sure to include at least a packet of edible-podded peas, also called sugar or snow peas. They are as tender as green beans, and are cooked the same way, but they taste like peas. You eat the whole pod. (To induce this flavor of the pod when cooking your other peas, snap off the end of each half of 10 or 15 pods and strip off the inside membrane, which is the coarse one. Cook in the pot with the peas.) The sugar peas to grow are Dwarf Gray Sugar, Burpee's Sweetpod or Super Sweetpod, which have thick succulent pods.

Pea seed is bulky, so you need a pound to sow 100 feet. A packet will only sow 15 feet, hardly enough to get more than a few pickings. It is advisable not to plant in succession, but to plant several kinds all at once which mature at different dates. The reason for this is that peas need cool weather to germinate well. If, however, it stays cool in your climate to the end of May, try some succession plants, too.

For a midget edible-podded pea, use Dwarf Gray Sugar, whose pods grow to 2½ inches and mature in 65 days. The plants only grow 2 to 2½ feet high, and are suitable for window boxes, roof gardening, and patio pots. After removing the dust, eat them, pods and all, as soon as the little peas can be seen inside.

For companionate planting, try carrots in the next row, or even between the plants in a row of peas. Onions and parsnips go well between plants, too, and so does lettuce. All will benefit by the strengths they will derive from root association and possible interchange below the soil of an antibiotic protection or some sort of growth-promoting substance, whatever it may be. They may even work to repel pests from each other.

If you keep your peas picked as they mature, you may well get a yield of a bushel for each 100 feet of row. Burpee's Sweetpod, if the soil, composting, and season are favorable, might yield two bushels per 100 feet of row.

A very good way to serve the first young peas is to steam them for a few minutes over rapidly boiling water and then, when they are tender, to drop them back into the warm pot after you have emptied out the water. Add butter and some heavy cream, and let it heat through. Serve in warmed little bowls, and eat with a spoon. Served this way, the peas are not only delicious, but the warm liquid keeps them hot longer. Plain boiled peas dumped on a cold plate have chilled before you get them to the table.

An even better way, we think, is to add raw young peas to the early summer salads we make from young lettuce, young radishes, and young raw spinach leaves. We use a very light soy oil and white wine salad dressing with this, and only a little salt for seasoning. Anything stronger would mask the flavors. Save the pods and make pea pod soup, with cream and two sprigs of mint.

Green peas are a very fine source of vitamin E (2.10 milligrams per cup). There are only three vegetables with a higher count—navy beans, sweet potato, and turnip greens. Peas are quite high in all the B vitamins, vitamin A (680 IU per cup as compared with 3,500 in broccoli, 1,110 in tomatoes, or 9,500 in turnip greens), and vitamin C (20 milligrams, comparable to asparagus or baked white potatoes, but only a tenth of what you'd get in turnip greens, mustard greens, or Brussels sprouts).

pepper

capsicum

Sweet or hot. Grow both, one for food and one for hotness and the excellent pest-repellent spray you can make out of hot pepper in the blender. But since they take more than 100 days to come to fruit, it is unwise to include them in your garden plans unless you are willing to buy the plants or start them indoors or in the hot bed. A very prolific, nicely fleshed sweet pepper is King of the North, and another is Burpee's Sunnybrook. One of the very earliest (60 days after 8 or 9 weeks indoors) is Merrimack Wonder, good to use in northern zones; another is a Japanese variety called Canape. Three varieties recommended as resistant to the mosaic virus disease are Yolo Wonder, Bell Boy Hybrid, and Keystone Resistant Giant.

For hot peppers to use in pickles and for an excellent spray, you have your choice of Hungarian Wax, which turns from yellow to red; Rumanian Wax, with sweet flesh, but hot ribs; Hot Portugal, the earliest; and either Tabasco or Large Red Cherry, very, very hot, and long-seasoned (85 to 95 days).

For those seeking sources of vitamin C, peppers are good, with more in the red than in the green. They have *pepper . . .* from 120 to 180 milligrams in each average-sized fruit. Also for vitamin A: 700 to 3,000 International Units; for thiamin: 30 to 70 milligrams; and for phosphorus: 28 milligrams. Harvest some green; wait until others mature and turn red.

Eat the pepper fresh. If you wait until it is limp, just about all the vitamin C has gone out of it, and much of the other food value. But fresh or stale, the calorie count is only 25, as careful weight watchers have already found out. As usual, the nutrients are more plentiful in the uncooked vegetable, but you can get tired of raw pepper after a while. Then a stuffed pepper is a welcome change. These are often baked, but simmered in a covered casserole they are just as good, and more mellow. Try a stuffing of ground beef or veal and ham, mixed with the yolks of eggs. Supplement or substitute brown rice and soybean grits, if you prefer.

potatoes

solanum tuberosum

If your garden is big enough, you want to grow potatoes. Even if it isn't, you want to grow potatoes; and since I've grown and harvested one in the kitchen window, I do not hesitate to recommend them for roof gardens, window boxes, and patio pot plants as well as for the garden. I know sweet potatoes make good house plants, too, but I doubt the harvest.

The organic gardener who mulches deeply just lays each whole or cut piece of potato right in the mulch, and waits for the new plant to grow. Sometimes this mulch is hay. Sometimes it is the last autumn's leaves piled up 3 feet high and left to weather down over the winter. You can also put seed potatoes on leaves and then cover them with hay. On sprouting, the roots seek the earth below and the plant leaves seek the light; the tubers which constitute the crop grow right in the mulch and rarely get scab or other soil-borne diseases. (Scab is a fungus disease that makes ugly brown spots on the skin of the tuber.)

For anyone preferring to plant in the soil, choose a well-drained, fine-loam area, somewhat sandy if possible, and dig a good trench 10 inches wide and 5 inches deep. Old, matured compost mixed with dried, old cow or sheep manure can be dug in during the previous autumn or a week before planting. But it must be old; fresh manures will injure the potatoes. Since they like acid soil, with a pH of 5 to 5.5, do not use lime or anything that will develop alkalinity and thus encourage scab. If your soil is not slightly acid, it is certainly best to plant potatoes in the mulch.

For 100 feet of row use about a half peck of potatoes, cutting them into four to six pieces. For good economy you can save eyes from the potatoes you eat in the spring. Cut them out with a small piece of flesh attached, wrap them in paper or plastic, and store in the refrigerator until planting time. Ordered seed potatoes should also be kept in a very cold place until planted.

Plant both early and late potatoes. For early ones, put them in as soon as all danger of frost has gone; for late, plant until the end of May. A few can be poked in during June, to see what happens. The average yield is supposed to be about three bushels per 100 feet of row. With careful practice and good composting you can double this. If you aim for a crop of ten bushels for a family of four, plant two or three rows, depending on your degree of fertilizer and mulch.

You know before you start that if there is one vegetable you thoroughly enjoy, it is a small, new potato, fresh from the garden, boiled whole in its jacket, and dribbled with butter. If you haven't yet had this experience, I can assure you there is a taste bliss ahead. When the flower buds show, begin to poke into the soil to find these little potatoes. You will be glad you did. Harvest the bigger ones later when the tops die down, and after the early ones are dug, use the rows for late planting of beans, kale, or turnips to get double use of your space. The late potatoes will provide space only for a cover crop.

This wonderful New World plant (from Peru) is a member of the *Solanum* family, which also includes the tomato, eggplant, and nightshade—as you can tell from the similarities of flowers and leaves. The plants and fruits of the vegetables we eat have little of the poisonous substance solanin—except in the early sprouts and growing tips. Avoid growing tomatoes, potatoes, and eggplants close together because as members of the same family they attract the same diseases and pests. Also avoid using the same ground in successive years. Plant a legume instead.

For varieties to choose, the early potato called Irish Cobbler is very reliable. For later ones, Katahdin and Kennebec are good. Sequoia is resistant to leafhopper; Kennebec, Saco, and Essex to late blight.

For pests, try some of the following: hose off aphids, and sprinkle the plants with homemade onion spray; pick off Colorado beetles and egg cases from the undersides of leaves; use homemade basil spray and mulch deeply for potato bugs—the mulch will thwart their climbing when they *potatoes . . .* emerge from the soil; sprinkle bran to stop beetles; introduce praying mantises or ladybugs; introduce milkweed as a trap plant for leafhoppers; wrap plants in plastic net for leafhoppers; and use the resistant strains of potatoes. If you have barn swallows, they can eat a thousand potato beetles in a few hours. Some people use poisons like pyrethrum (getting it from the pet shop or veterinarian), nicotine, or ryania. Unless you have a big patch with an unusual attack, you probably will not need to resort to such sprays. Another possibility against the Colorado potato beetle is *Bacillus thüringiensis,* cleared for this use by the United States Department of Agriculture. Twenty-three insects are affected by this disease. If you think of using it, consider the fact that there are at least that many species of birds who eat insects and might depend on some of them. In the South, keep all grass out of the patch, for potato wire worms breed there. Also in the South precede the potato crop with a crop of soybeans or some other legume on the same soil. Keep the land well cultivated both years.

To protect your plants from blight, keep everything clean and dry. If blight gets into your potatoes during the cool, humid days of fall, leave them in the ground for two extra weeks so the spores will all die down before they get a chance to float out from the stem to the tuber when you disturb the plant and stir up the spores while digging. Keep all tools, shoes, paws, and hoofs clean, and burn all culled plants. Never let plants get wounded in hot weather. Never use soil or stock that you know was recently infested.

Store potatoes in a safe place with a temperature around 40 degrees. We put ours in bushel baskets in the cellar. The mice like that. A better plan, which we have not used, is said to be to swaddle them in leaves and dry litter in a big hole in the ground. But wouldn't the mice like that, too? Probably the best strategy is to build a screened cage in a cool place, or even to screen over your bushel baskets.

Does anyone need a reminder any more that the skins of potatoes should be eaten? Peeled, boiled potatoes have lost 47 to 50 percent of their vitamin C by the time they are cooked. And if you mash them, 10 percent more gets away from you. In addition, if you have soaked them, you have lost some of the vitamin B group, too. There are so many excellent recipes for potatoes it seems silly to add any cooking hints aside from those warnings. I'm still not able to eat raw potatoes because I have a notion they put my teeth on edge, but grated potatoes, fried very quickly immediately after grating, can be quite good. I also slice them very thin (with the skins on) and bake the slices in a 350° oven in a little oil until crispy. A very easy recipe.

For variations on old favorites: instead of using milk and butter for mashed potatoes, use yogurt and chives, or yogurt and thyme, watercress, sorrel, or mustard leaves. For leftovers, make patties to fry or broil, and add wheat germ, leftover soy grits, mashed wheat berries, or ground nuts to make the potato stiffer. Use half a cup of the addition to two cups of leftover mashed potato. Coat each patty with wheat germ or soy flour, and add some ground marjoram or rosemary.

According to *Cooking with Astrology,* by Sydney Omarr and Mike Roy, you are not likely to enjoy potatoes very much if you were born under one of the first three signs of the zodiac. But if you are a Cancer, you like duchesse potatoes when flavored with mace or nutmeg and dropped in dollops on a cookie sheet. You like potato patties and fritters if you are a Virgo, made as in the suggestions above, but with egg yolk and beaten egg white added to make a very soft batter. If you are a Scorpio, you like molded potatoes made by packing slices of potatoes in a mold with plenty of butter and covered with foil or parchment before baking until very tender. Plain or fancy potatoes appeal to the Capricorn lover of the Good Earth, but if you are a Pisces, you probably prefer your potatoes in chowders or fish stews, although a casserole of potatoes, cottage cheese, and sour cream would suit you well, too.

Sometimes go back to the plain baked potato. Rub it first with a little butter or vegetable oil after washing off the bits of soil, and put it in a hot oven—400° or even 450°. Do not let it overcook, and prick it with a fork to let in a little air when done so it won't get soggy. In this form it preserves more of its nutrients than in any other. It is not fattening unless you douse it with butter or gravy. If medium in size it will provide you 33 milligrams of vitamin C, about the same as in a large tomato, and 1.5 milligrams of iron, about the same as in a large egg. If you go to the trouble of peeling, soaking, boiling, and ricing this potato, you will have thrown away 83 percent of the iron and absolutely all of the vitamin C. To save the vitamin C otherwise lost when a potato is cut open, chill it thoroughly first. And if you do have to cut a potato ahead of cooking time, put the pieces immediately in an airtight bag and return to the refrigerator.

pumpkin

cucurbita pepo

Halloween, pumpkin pies, and an occasional stewed pumpkin when you have no winter squash are the usual reasons for growing this big husky plant. Shredded raw pumpkin to add to a salad or to fry quickly until crisp is another kind of pumpkin you may not yet know about, but when you do, you will be glad of the taste it offers.

Culture for the pumpkin vine is usually the same as for squash, in hills spaced four feet apart or in a row thinned eventually to about that distance. Interplanted with corn, pumpkin will do fairly well, seeming to be helped by the shade of the corn plants. Otherwise plant on the edge of the garden so they'll run onto the grass. A packet of seeds will plant five or six hills, and with fertilizer and good watering this amount might well be enough for all your needs. The new bush pumpkin, which takes less room than the older varieties, is called Cinderella, and it is quite a surprise to see pumpkins grow as summer squashes do on a squatty bush. This variety matures in 102 days, as compared to the 115 days for the old varieties such as Connecticut Field. The fruits weigh up to 10 pounds, as compared to 20 pounds for Connecticut Field. For an in-between pumpkin, very round and stylish, try Youngs' Beauty.

Any of these pumpkins are among the great bargains of the garden. You can get a packet of seed for 35 cents this year, and from those seeds you can get several hundred pounds of food. What's more, they store very well with almost no loss of food value; and don't forget the very nutritious seeds, easily dried in the sun or in a warm, dry kitchen. They are as good as sunflower seeds and almost as nutritious. You get from each serving of baked pumpkin 3,400 International Units of vitamin A, 37 milligrams of vitamin C, and big provisions of minerals. It is a mystery why Americans shun this excellent food. If it tastes flat to you, add spice or herbs.

radishes

raphanus sativas

Easiest vegetable of all. Radishes are easy to plant, for the seeds are large and manageable. The red ones come up and are ready to eat in 21 days, white icicle varieties in 30 days. Even the winter radish, Round Black Spanish, is ready in 55 days after a late sowing in July or August. The tops are edible, too, though rather ticklish when raw.

The varieties we like for red radishes are the

quick-growing Red Boy and a very round, smooth radish called Cherry Belle. Both are solid and pleasant, and if you slice some very thin and fry them two minutes in oil, they are soft and mild and taste a little bit like turnip. These two varieties have short tops and can be grown indoors as well as outdoors. Try them in the kitchen or on the terrace. A slower-growing variety, Champion, has tops that sometimes get quite large. The tops are tougher, but they make a cooked green if you want variation.

For white radishes, there is one 25-day variety called Burpee White and the 30-day White Icicle with roots that grow up to 5 inches. A midseason one is the very satisfactory White Strassburg, a 40-day radish. One which has the virtue of holding for several weeks in the ground without becoming pithy as most others do is called All Seasons.

You won't have to thin radishes if you space them well while planting or if you intermix them with carrot seeds or herbs when you plant. Just pull them as they mature, and eat them up. You must thin them wherever they grow too thick from careless planting, for then no roots develop well, and all you have is greens. One excellent place to plant radishes is among the hills of cucumbers as companion plants.

Because they are pungent, radishes are bothered very little by insects, and actually drive away cucumber beetles from squash and cukes. But they do occasionally get root maggot, especially near corn and cabbage. In fact, radishes are used as a trap crop to attract maggots away from corn and cabbage. The best deterrents are never to plant in soil that was infested the previous year, always to give the soil good deep preparation with plenty of compost, and start out with a large dose of wood ashes.

Cut and soaked radish roses may be pretty, but you lose a good portion of their nutriment if you treat them this way. Cut off the tough lower root, and some of the top, and put them in your salad or on your snack dish whole. Leaves fried in hot soy oil or butter will be cooked in two minutes or less. Drain them on a piece of bread. Since they dry out very rapidly, they are fragile and break easily. Best used as a garnish.

rhubarb

rheum rhaponticum

This useful perennial was once found only in apothecaries' gardens, but now is grown alongside many home vegetable gardens for use as sauce and in pies and conserves. Buy six

Relatively Pestless Leafy Vegetables

or eight roots and set them out in a rich, well-prepared bed. Do not harvest the first year, and pull out only a few stalks the second year. You can also start rhubarb from seeds and again plan not to harvest until the second year. The varieties to choose are MacDonald, Valentine, and Victoria. MacDonald has big red stalks, of a good flavor. Rhubarb can be divided every eight or ten years by digging up the plants and cutting the root in several pieces before replanting.

rhubarb . . .

When the plant comes to flower, cut out that stalk. It uses up food that might better go to the roots to feed the new leaves the following spring. In the fall give the plants a good spread of extra mulch, and manure them well. If you want to dig some roots in order to have winter rhubarb, take them up in late fall and store them in sand. Leave outdoors until after the first hard frost, and then put them in the cellar. Never permit dock to grow in the neighborhood of your rhubarb, for it will entice the rhubarb curculio pest to come into the area. And never eat the leaves of rhubarb; they will make you sick.

Every spring I make several batches of Aunt Mary's Rhubarb Conserve, invented for the White Turkey Restaurant. Take 6 pounds of rhubarb stems, cut to 1-inch lengths, and combine with 6 pounds sugar, 2 pounds seedless raisins, cut up, 1 pound walnut meats, cut up, 4 oranges, cut up, skin and all. Mix well and boil for an hour. Put in sterilized jars and cover well.

Do not peel or scrape rhubarb stems when you prepare them for conserve, pie, or sauce. That's just throwing away nutrients to your compost heap which you might as well have first.

rosa rugosa

Though you may not want to introduce this species of hedge rose into your vegetable garden, consider it seriously for other parts of your place for the sake of its very nutritious rose hips or seed pods, an excellent source of vitamin C. Anywhere you want an accent or a border plant or a hedge, it will probably do well. This rose can be ordered from most nurseries that specialize in ornamental shrubs, roses, and decorative hedges, though it is sometimes not advertised by name. It may be called everlasting hedge rose or something like that, but when you write refer to it as *Rosa rugosa* anyway. Stern's Nurseries, for example, do call it by name. It has lots of unspectacular crimson-pink flowers, from June until frost. Cultivation is easy, with the only requirements being a good, rich soil. These roses are relatively pest-free, not nearly so pestered as the hybrid tea roses. Pick off the few rose bugs you may see. Harvest when first fully plump—unless you want to leave them for the birds.

For food value, note these interesting comparisons: vitamin C, 2,000 International Units in rose hips as compared to 50 in an equivalent amount of orange fruit or juice. Vitamin A, 5,000 IU, compared to 200 in oranges. Protein, 1.2 percent and carbohydrate 17 percent as compared to .9 percent and 11.2 percent in oranges. To get the full vitamin C benefit from oranges they must be fresh. In roses, the vitamin C persists, even in dried, powdered form, though the hips should be chilled as soon as gathered to prevent rapid deterioration by enzyme action.

When they are cool, and ready to prepare, remove the stem end and the blossom end of the seed pod, and for each cup of hips have ready 1½ cups of boiling water. Simmer for 15 minutes, and let stand in a glass or pottery vessel for 24 hours. Strain and bring the extracted juice to a full rolling boil. Add 2 tablespoons of lemon juice for each pint of rose hip juice, put it up in jars and seal. Never use aluminum or copper for storing or cooking these fruits, any more than you would for tomatoes, rhubarb, or other acid foods. Then use the juice to make cobblers, jellies, candies, puddings, and to spruce up drinks.

For rose hip jelly, add a box of commercial pectin, or homemade apple pectin, to a quart of juice, and bring to a boil. Add an equal amount of sugar, bring to a high boil for a minute and then skim and put in sterilized jars.

To dry rose hips, split them in half, spread pieces on a cookie sheet, and toast them slowly in a 200° oven until they get crisp. Do not remove the seeds. The pulp from making juice or jelly can be dried also. It makes quite good tea, but eat the grounds, too, to get all that vitamin C. Dried rose materials keep well, if kept clean and dry.

Rose hips from any variety of rose can also be used in all these ways.

rosemary See *Herbs,* p. 170.

rutabaga See *Turnips,* p. 205.

sage See *Herbs,* p. 170.

salsify

(oyster plant) tragopogon porrifolius

This unusual and highly nutritious plant can be a good source of food over the winter, for it is easily stored. It is a root vegetable, and takes about 120 days to mature. Part of the crop may be left in the ground as you leave carrots and parsnips to be ready to dig in the spring, or during the winter if you protect them so you can get at them. (See p. 138.) They grow to about 8 inches in length and have a diameter of an inch to an inch and a half. They do taste oddly like oysters, especially when well buttered.

The variety to get is Mammoth Sandwich Island, and sow the seeds half an inch deep in rows 2 feet apart in May. Thin them later to 3 feet apart for the individual plants. They grow best on light, rich well-composted soil, and will do well if occasional side dressings of old manure are applied down the rows. Two ounces of seed will plant 100 feet of row, which will give you a big supply.

A very good stew with oyster plant is made by melting 4 tablespoons of butter in a skillet; in it lightly brown 1 cup of diced raw veal. Then add 1 sliced carrot, half a medium-sized onion, sliced, 1 tablespoon chopped parsley, 1 tablespoon tarragon, and 1 tablespoon marjoram. Next add 1 cup of hot chicken broth. (Soy milk powder and green drink may be substituted for the meat and broth.) Cover and simmer for 2 hours. Meanwhile clean and simmer ten oyster plants, and when nearly done, transfer them to the other pan, leaving them there long enough to gather up the flavors. In about half an hour add the juice of a lemon and 3 tablespoons butter cut in small pieces.

savory

See *Herbs,* p. 171.

soybeans

glycine max

Should be a staple in any organic garden. Though the soil and cultural needs are about the same as those for bush green beans, the maturing time can run from three to five months, so adjustments must be made if you live in a cold climate. For midseason varieties, you can plant any time between April 10 and June 30, if your last frost is on March 20 or thereabouts. If it is as late as April 20, however, you plant between May 15 and June 15, thus allowing for the earlier frost at the other end of the season. If your last freeze averages any later than April 30, you are on your own—you may succeed in growing soybeans and you may not.

The zone having an average last freeze around April 30 is a big one, for it not only stretches across the country in the moderate belt, but also comes up the Atlantic coast from Virginia to eastern Maine, with a few extra pockets on the Hudson River and around the Great Lakes. To extend the season you can use Hotkaps, large cold frames, or a row cover of heavy plastic. In northern areas grow Giant Green, which matures in 90 to 95 days, and avoid the longer-growing varieties such as Hokkaide and Imperial, which take up to 115 days. Bansei is a middle-season soybean, taking 95 to 100 days, and Kanrich from Burpee's takes 103 days. If you can harvest them, these midseason beans are versatile for eating green, frozen, or dried.

The soil you plant in should be well drained, fairly fertile, well limed, and well mulched. It should also be warm. To hasten warming, you can put out large plastic bags of water along the rows where you are going to plant. They will accumulate the heat of the sun, and hold at least some of the warmth overnight. They are so big they cool off slowly.

A packet of soybeans will plant 25 feet of row. Put the seed in a furrow about 2 inches deep, and when the seedlings are 2 or 3 inches high, thin them to stand 4 to 6 inches apart. You may have a problem with weeds.

The plants grow very slowly at first, and if the weeds get a start on them, the little soybeans do not have much chance in the struggle for root space, water, nutrients, and light. Then, if you start yanking out the weeds right next to the very small plants, their root hairs are disturbed, even ruined. If you leave the weeds to devour the available nutrients, your beans will be spindly and lower in nutritional value than otherwise. So the thing to do is to get the weeds out before you plant. Cultivate the soil, let the weeds grow up, and then cultivate again and immediately mulch. When it is time to plant, push the seeds down in through the mulch, leaving an opening big enough to let in the light and air the small seedling will need. The mulch will help preserve moisture, too, of course.

You may have a problem with Japanese beetles. Plant white geraniums among the beans, put around milky spore disease, pick off beetles by hand, and let the house sparrows come in to eat them. In the south, for velvet bean caterpillar, you may have to bring in *Bacillus thüringiensis,* which you can get under the trade name Thuricide or Biotrol.

Keep experimenting to see what method works best

for you for outwitting the warm-weather, long-growing season needed by this most excellent of all sources of vegetable protein. If you can grow soybeans successfully, you will have added bountifully to your garden produce.

Harvest the green beans, as soon as mature; and let them stay on the vines a little longer for dried beans. When green soybeans are cooked 20 or 30 minutes in a little water or milk, they taste like a nutty version of lima beans. But they never get mushy. To make them easier to shell, blanch or steam the whole bean for 4 or 5 minutes, and then you can snap them in two and squeeze out the beans rather easily. They will stay springy, so do not overcook them; if young and just picked, they might be ready in 15 minutes.

With leftovers make a succotash, soup, casserole, vegetarian chili, or salad. One soup starts with ¾ cup of grated carrots and 5 tablespoons minced onion, sautéed for 5 minutes in 1 tablespoon of soy oil. Then add 2 cups of cooked soybeans, 4 cups of chicken stock or tomato juice, and 1 teaspoon of fresh marjoram, or whatever herb you prefer. The beans may be left whole or blended with two cups of the stock first.

Salad mixtures that are particularly good are green soybeans and cucumber with chives; or soybeans, celery, and onions. A high-protein fritter can be made with 2 cups of mashed beans, 2 beaten eggs, and 2 tablespoons of grated onion, fried on both sides until brown on a hot greased griddle. You will be getting a protein content of 12.5 percent. Eggs have 14 percent, veal has 16 percent, steak has 19 percent, and chicken 22 percent protein.

If you mature and dry soybeans, the protein content goes up to 35 to 45 percent. Because dry soybeans are so high in protein, they need a lot of soaking and cooking to make them tender. When ground to grits, they cook a lot faster.

For whole dried beans, soak them overnight, then put them in the freezer for several hours to make them even tenderer. An hour before using, remove and cook them in a pressure cooker for 45 minutes (or boil them for three hours) and serve plain or with gravy. Make the gravy by stirring 1 tablespoon butter and 1 tablespoon whole-grain wheat flour to the liquid drained off after boiling. For additional flavor, some sautéed carrot shreds, finely chopped shallots or onions, and celery leaves may be added.

Cold dried cooked soybeans are good in a brown rice and green pepper salad, and, either whole or ground up, they are valuable additions to many kinds of casseroles. They will provide the protein for a vegetable loaf and for soufflés, croquettes, and stuffed tomatoes or peppers. They make good baked beans, with molasses or honey and a piece of onion. A fine dip can be made by mixing 2 cloves of garlic, minced, 2 tablespoons mixed herbs, ½ cup softened butter, and about 1½ cups cooked ground beans. Thin with mayonnaise if necessary. Can be made in blender.

Soybean sprouts are easy to make, and very versatile and nutritious. They are excellent in salads, for tempura, folded into scrambled eggs or omelet, in mashed potatoes, or stewed in a little water, milk, or tomato juice.

To make the sprouts, soak a cupful of dry soybeans overnight in cold water in a large glass container. Flush it all out, put on a cloth cover, and keep in the dark. Keep flushing the beans four to six times a day to keep down fungal growth. In four or five days you will have sprouts two inches long, ready to use. Bring them to the sunlight for a few hours to green up, then put them right in the refrigerator, so their plentiful vitamin C will not be lost.

Another delicious way to eat soybeans is as salted nuts. After washing dry soybeans and soaking them overnight, drain them and spread them out to dry. When thoroughly dry, deep-fry them for 8 or 9 minutes in a good fat heated to 350°. Or bake them in oil for half an hour at 350°. Drain them on absorbent paper or pieces of bread, and salt them while they are still warm.

Soybeans can also make fine-tasting grits blended or ground at home from dried beans, or in the food mill at your natural foods store. When the soybeans are ground, the protein content goes up to 56 percent, the same percentage as low-fat soy flour (which has had the bean fat taken out).

Soybean flour is made as low fat or high fat. It is yellowish, fluffy, and a pleasure to work with, though it must be mixed with other flours, for it has no gluten. Its protein makes it like a substitute for powdered eggs or powdered milk; and it has triple the protein percentage.

You can make soy milk, too. Soak dried beans for 12 to 24 hours, changing the water frequently. Grind the beans to a fine paste in a food chopper, using the finest blade. Keep adding water—up to three times the amount of beans you have. Boil to a high foam for an hour, and strain through cheesecloth. Or run the beans through the chopper, add 6 cups of water to 1 cup of beans, simmer for 15 minutes, and strain. Or make in the blender. Use it as you would any milk. If you want it richer, blend in a cup of soy oil and 2 tablespoons of honey.

Soy Ice Cream: Put in a blender 2½ cups soy milk, ¼ cup corn oil or soy oil, ½ cup honey, and a pinch of salt. Blend well, and add ½ cup chopped fruit and ½ cup nuts. Freeze in a hand freezer. (Strawberry ice cream can be made with the basic recipe, omitting the nuts and increasing the fruit to 1½ cups. For more body, 1 cup of soy milk powder may be added, especially if the strawberries are very large, fresh, and juicy.)

There is one more soy product, soy cheese or *tofu,* which is used as cheese, sandwich spread, in desserts, and again as a substitute for meat or fish. It can also be frozen, in or out of water. Make it by allowing soy milk to sour. Put a cup or so in a shallow dish in a warm place, and leave until thickened. Cut in pieces, cover with water, and bring to the boiling point. Then strain and squeeze the curds dry before placing in cold water for storage. The best way to eat it is in sukiyaki.

spinach

spinacia oleracea

A crop that goes by so fast that you sometimes wonder whether it is worth growing or not. However, if you have the room, plans for replanting, and someone to help you blanch and freeze it when the whole crop begins to bolt and has to be used, go ahead and plant it.

We start adding the tender outer leaves to our salads as soon as they appear and keep snipping off and adding leaves as the plants grow. Then we have a meal or two of steamed spinach before the day the plants begin to shoot up and bolt.

Since spinach is a cool-weather plant, succession plantings can be made throughout April and into May until the earth gets too warm. You can plant it very early, for it doesn't mind the frost. It is also possible to plant again in late August so it will come again in the fall. Varieties to try, aside from old standards like Viking, are new so-called long-standing varieties like Bloomsdale and Winter Bloomsdale, which can endure more heat than Viking. In addition there is the New Zealand spinach (not really a spinach), which is heat-resistant and is called everlasting because when you pick it, it grows again. Try it. You might like its watery taste.

For spinaches, use very rich, nitrogenous soil and

spinach . . . compost, with a pH of 6 to 7. Sow in rows 12 to 18 inches apart and cover the seed with half an inch of soil. Thin the plants to 3 inches apart when they get to be 3 inches high. For an ounce of seeds, you ought to get three bushels or fifty pounds in a 100-foot row. Remember, however, that the waysides and your own garden are full of other greens to cook the same way you do spinach (see *Day Lily,* p. 152, and *Weeds,* p. 208).

On the young spinach leaves there may be a few pests to bother you. One is the spinach leaf miner, which goes into the veins, curls them up, and makes the plants look horrid. Pull up any affected plants you see and bury them in the compost heap. Rout out all weed hosts you notice, and bury them, too. A few blights like fusarium wilt can get into the part of the garden where you grow spinach. If it does, allow at least a three-year cycle before you plant spinach there again. Also switch to the Virginia Blight-Resistant variety.

For a hearty spinach dish, try it with cheese. Combine two cups cooked, chopped spinach with 2 cups cottage cheese, ½ cup grated cheddar or Parmesan, 1 teaspoon salt, and 2 eggs. Mix all together until well blended. Put into a 1½-quart mold or casserole and cook for half an hour at 350°. When still hot and ready to bring to the table, sprinkle on more cheese.

squash

cucurbita

A cucurbit, which makes it cousin to cucumber, melons, and cantaloupes. Plant several kinds: for summer use, yellow summer squash and zucchini, cocozelles and straightnecks; for winter use, acorn, blue, Hubbard and the excellent new buttercup squashes.

If you live where you need to irrigate, you can build a ditch or moat of about four feet in diameter around each hill of squash and water your plants in that. The roots will get to it, and the water will be held from eroding away.

Have the soil where you plant very loose. If it is a little acid, the squash won't mind. When possible, plant at the edge of the garden but on a side away from the pumpkins so that when the vines crawl out, they will not get into your other vegetables. You can also pinch back the wandering tips, to keep them under control. For amounts, a quarter of an ounce of seed will sow 50 feet, with one or two yards between the plants when they become mature.

squash . . .

Plant summer squash in April or May; winter squash in May or June. Choose bush varieties to save space.

Put in more of the winter kinds than the summer because they are easily stored and will last you over the winter. Pick these just before frost. Wash, dry, and rub them lightly with soy oil or corn oil, and put in a cool, dark place. Do not worry about any leftover seeds you may have after you plant, for they will stay viable for a year or so. Or you can eat what you do not plant. Do not plant seeds from your hybrid cucurbits; you'd get a strange crop. (Ditto for hybrid corn.)

All squashes are heavy users of nutrients, so dig big deep trenches, fill them up with several inches of well-rotted or dry manure, then compost and add fine sandy loam before you plant. A good soaking whenever needed will also help growth, and a careful inspection every morning for cucumber beetles or squash bugs is advisable during the period of early growth.

If you get any such pests, spread wood ashes, or make a spray of wood ashes and water to use on the leaves. When there is a hot spell in June, use wood ashes anyway, as a deterrent. You should also interplant calendulas and marigolds as a deterrent to cucumber beetles, and try out an onion-and-water or garlic spray if these pests are persistent. Also pick them off and toss into kerosene. This fine, nutritious vegetable is good raw, boiled, sautéed, or baked.

sumac

rhus glabra

Scarlet sumac or staghorn sumac is a tall shrub found along many roadsides. Its fruit ripens to red plumes in the summer, and its leaves turn bright red in the fall. The malic acid in the fruit makes a delicious tart drink and jelly. Cover with water and crush the horns, or fruits, with a potato masher, and strain the juice through a fine-mesh cloth. Very fine mesh. There are prickles to get out. Only fresh young fruits are good. Stale ones are bitter. (Euell Gibbons' son does the crushing in a washing machine.) Don't confuse with the white-berried swamp sumac, which is poisonous.

sunflowers

helianthus

Several varieties. A good crop to grow, and what you don't harvest for yourself or for winter feed for the birds, you can leave right there in the flower heads and the birds will come to harvest it themselves. It is pleasant to see the first goldfinches fly in to get their fill.

sunflowers . . . **Sunflowers** do best in a deep, rich soil, slightly moist, but will survive in many different kinds of soil. The seeds can be planted about half an inch deep and about 12 inches apart, or more if you grow the biggest varieties. The seeds are very high in protein, calcium, phosphorus, and several of the B vitamin group, and have long been relished as a snack. The oil of the seed is a source of both linoleic acid and lecithin. For anyone planning to grow large quantities, you should add lime to adjust the soil to a pH over 6, say 6 to 8, and apply plenty of manure, rock phosphate, and granite dust with the lime. At harvest time, begin to cut the heads off as soon as the outer seeds get ripe—or as soon as the birds start coming in to peck out those outer seeds. Hang the heads in a warm airy place, and they will continue to ripen for a while. When ripe, rub out the seeds and store in airtight jars.

Varieties to consider are the giant 12-foot perennial sunflower, *H. giganteus; H. annuus,* the common annual sunflower, which can grow up to 12 feet; also a 5-foot one called *H. decapetalus,* a perennial. When you order seeds, ask for Mammoth; it produces lots of seeds on good, firm heads.

It is all right to plant sunflower seeds in a row, but it is also a good plan to scatter them around the garden and wherever else around the place you can find a good bit of soil to put them in. They may need staking when they get big. In the garden they cause a lot of shade at this size. The big roots compete for nutrients, so add compost.

If you raise your own sunflowers, and especially if you also raise your own soybeans, and nuts, you have the beginnings of making your own Granola: To a cup of hulled sunflower seeds add a cup of chopped nuts, 4 cups mixed rolled oats and rolled wheat, 1 cup wheat germ, 1 cup ground roasted soybeans, and ½ cup sesame seeds. Heat ½ cup honey, ½ cup soy or safflower oil, and 2 teaspoons vanilla. Mix this well with the grains and nuts and spread out on a cookie sheet oiled with soy oil. Bake for 30 minutes at 375°, and stir occasionally. A cup of shredded coconut can be added before cooking, if desired.

tarragon See *Herbs,* p. 171.

thyme See *Herbs,* p. 172.

tomatoes

lycopersicon esculentum

A delicious, nutritious, and very inviting vegetable when it comes warm and ripe straight from the vine. Any other tomatoes, even the ones called "hothouse" from the market, have a completely different taste from those thoroughly ripened in the garden.

Tomatoes are a long-seasoned fruit, so it is customary to start them indoors, though volunteers from the previous year and seeds planted very early outdoors, if they can make it, produce husky heavy-bearing plants in the fall. It is best always to plant tomatoes in ground not recently used to grow tomatoes, potatoes, eggplants, okra, or peppers. Be sure the place you select will have plenty of ventilation, and plan to provide quantities of nutrients, compost, mulch, and water for your maturing plants. But water only early in the day because tomato leaves and stems should be kept as dry as possible. Do not let dogs run near the plants when they are wet or damp. Tomatoes wound easily, and then diseases get into the wounds.

This is one plant that you do not toss on the compost pile unless you are sure it is absolutely healthy. Even its seeds will come through the very hottest of piles. And do not smoke or chew tobacco in among the plants, or handle them after you have filled a pipe and touched tobacco. Tomatoes cannot endure the results of exposure to tobacco, which causes tobacco mosaic virus.

To safeguard yourself somewhat from these hazards get pest-resistant strains, and some time try out preplanted seeds from such seedsmen as Burpee. They will provide a verticillium- and fusarium-resistant strain called Burpee's VF. Also try Burpee's Big Boy Giant Hybrid and Big Early Hybrid in a setup called a Fertl-House. Harris has a resistant strain called New Yorker. Stokes has it, too, and another called Maritimer. If you have someone else grow your tomato plants for you, ask for resistant strains of this sort. Remember that it takes at least 6 to 8 weeks to get a tomato plant to the size proper to set out in the ground, so order early. But do not press the season; in most areas the end of May is plenty of time to set out tomato plants. People ruin them by putting them out too early. We wait to put ours out in June, and begin to harvest in mid-August.

Some other recommended varieties include Globemaster, Manalucie, Rutgers, Glamour, and Supersonic. And there is a good, early variety called Fireball. For a change, grow some orange-colored tomatoes, mild and very high in vitamins C and A, such as Sunray; and some pink ones, such as Ponderosa, Livingstone Globe, Crackproof Pink, and Early Detroit. For a midget tomato try a Dwarf Champion, a tree-like variety that is very well adapted to patio pots and window boxes or small raised beds. Other small tomatoes include the yellow ones—Yellow Pear and Yellow Plum—and the red cherry varieties—late, early, or small.

The food value of tomatoes is very high. They can supply in an average serving 1,000 International Units of vitamin A and 13 to 30 milligrams of vitamin C, plentiful portions of eight of the B vitamin group, and minerals including cobalt. For the best supply of vitamin C grow tomatoes on pole supports so the fruits get lots of sun. Any outdoor-grown tomatoes have twice as much vitamin C as hothouse ones. Do not ripen picked green tomatoes on a windowsill in the sun. Keep in sunless full light.

You can cook nearly everything with tomatoes—sauces, soups, spaghetti, stews, juices, and even cake. Toward the end of the season I begin to bring in some green tomatoes to slice and cover with salt and flour, then fry in butter. They have a delicious tart flavor. I also make green tomato relish:

Pick enough to make two quarts of cut-up pieces. Cover the pieces with $\frac{2}{3}$ cup of salt and let them stand for a day. Add $1\frac{1}{2}$ teaspoons each of pepper, mustard, cinnamon, cloves, and allspice, $\frac{1}{3}$ cup of white mustard seed, 2 or 3 chopped onions, and a quart of vinegar. Bring to the boil and cook for 15 minutes. Pack in sterilized jars. They are quite crispy, and go well with all kinds of baked beans and loaves.

The first gigantic plant I ever knew was a single tomato plant nurtured and tended so it produced enough fruit to satisfy the needs of a whole family. It was grown against a fence, and had a chance to spread out as it wished. The family tied each branch very carefully with soft strips that would nowadays be cut-off nylon stockings, I suspect. First it was given a huge 6-foot hole, half corncobs and half well-rotted but unleached manure. This was covered with fine loam and compost mixed to the right texture for receiving the young roots when the tomato was planted. Because the location was a sandy seaside spot, the plant was also supplied with a length of hose that went down to the nutrient area, and the tomato was watered every day. Since tomatoes are tolerant to a pH range from slightly acid to neutral, no liming was necessary. As you can imagine, this plant thrived. It was so healthy it was pest-repellent, and besides tomatoes have an insecticidal alkaloid somewhat like digitalin in their leaves. Any worms that arrived were picked off, and the leaves were given a protective spray once or twice of hot peppers blended with water with a drop of detergent in it.

turnips

brassica rapa

Sturdy vegetables, one variety small, one big called Swede or rutabaga. Well adapted for immediate use and for storing into the winter. Turnips can be grown both early in the spring and in the fall, but are considered much better if grown in the fall. This means that you should get seeds in the ground in late July in the Northeast. Do not plant in the spring, as they will bolt and get woody stalks. When they first come up, there is a tiny flea beetle which bothers the fresh, green growth. It can be controlled by a dusting of rotenone—but try wood ashes first.

Keep the rows a foot apart, and thin the plants to stand four inches apart. The recommended variety is Just Right, which will mature in 35 to 40 days. Its fast-growing greens give a wondrous supply of vitamins and minerals within a few weeks of planting. If you like their pungent taste, you ought to plan to eat lots of these greens. The large, smooth white roots are milder in flavor, and will be ready to harvest before frost if you plant the end of July. Another good variety to grow is Tokyo Market, which matures in 50 days; and a longer-growing kind is Purple Top White Globe, which takes 58 days. Since these last two are slower to bolt than the others, they can be planted in the spring. Plant some for the greens, anyway.

These greens will give you more calcium than any other—a stupendous 259 milligrams in an average serving. Next you get phosphorus—50 milligrams, surpassed by only six others; then potassium, 440 milligrams, surpassed by only five; vitamin A, surpassed by only two; and vitamin C, surpassed by one other vegetable, and that is parsley. If you happen to eat both turnip greens and beet greens in any quantity on the same day, you can run the danger of locking up the calcium intake in salts that form with the greens' oxalic acid. If you want to eat two or more greens in one day, combine your turnip greens with lettuce, dandelions, or one of the endives. The more you are willing to eat raw of both tops and root the better.

Turnips are light feeders, so should be rotated with such heavy feeders as kohlrabi, corn, or squash. Fertilize them with a moderate application of compost and a little rock phosphate and granite dust, never with heavy manures.

Plant scientists have found that turnip plants have some sort of insecticidal chemical compound in their systems which is deadly to aphids, spider mites, and house flies as well as beetles. Therefore it seems only sensible to use turnips as a repellent for those pests and interplant them with beans, for instance, which might be bothered by at least two of those pests.

The time to dig turnips is before they get big, woody, and bitter. In September, begin to pull away the soil and look at them, or pull one up to test it. You can leave them in the ground till frost, but not if they are going to get big and woody.

Hardiest for winter storage is the one big turnip that is supposed to be big, the rutabaga or Swede, *Brassica napobrassica.* These vegetables get so big they may seem coarse, but they can be very tender and rather spicy if grown and prepared properly. They are longer growing than turnips, so the seed should be sown in late June rather than in July, in rows about two feet apart instead of one foot apart. The two best varieties are Macomber, an old favorite with many because it stores well, and American Purple Top, or Burpee's version of this, Purple-Top Yellow. Two packets of seeds will sow 100 feet.

I think that some of the most interesting recipes for turnips come from soul food cookery. One recipe that has a very good pot liquor calls for a pound of salt pork, ham hocks, or fresh hog jowl, 3 pounds of turnip greens or young rutabaga greens, 4 cups of cold water, 1 tablespoon vinegar, ½ teaspoon crushed red pepper, ¼ teaspoon fresh-ground black pepper, 1 teaspoon salt, and at the end a garnish of sliced sweet onions and sliced hard-boiled eggs.

Cover the meat with cold water, add the seasonings, and bring to a boil. Then turn down heat and simmer for an hour. Add washed greens, discarding all yellow leaves and stems. Simmer one hour longer, but remove the cover during the last 15 minutes of cooking. When you drain the greens, reserve the pot liquor and meat. Chop the greens if you wish, adjust the seasoning, add a little vinegar if it suits your taste. Put the greens in a serving dish and add the meat and garnishes. Either serve the pot liquor separately in a pitcher or pour it over the greens in the dish, depending on your preference. Serve with corn bread, to be used to douse up the pot liquor.

Another variation of a recipe from *Tuesday Soul Food Cookbook* is Mixed Greens with Corn Bread Dumplings, a large recipe for 6 to 8 people. Strip the stalks of 2 pounds of turnip greens, 1 pound of mustard greens, and 1 pound of collards or spinach. Wash very quickly in cold water, to get the grit out, and put in a big (8-quart) pot. Add 2 quarts of boiling water, 1 pound of salt pork or bacon, and 1 chopped onion. Boil for half an hour before adding the greens. Add 2 tablespoons sugar and some salt if needed. Simmer it all, covered, for 2 hours. For dumplings to drop into the pot liquor, mix 1½ cups of water-ground white or yellow corn meal, 4 tablespoons flour, 1 tablespoon salt, and 1 tablespoon sugar. Add 1 cup of boiling water, stirring until the mixture is stiff. Wet your hands, make small balls of dough, and drop them in the pot of greens. Cover tightly while they simmer for 30 minutes.

These soul food recipes, though they do not follow the preferred natural foods methods of quick cooking, do preserve many minerals and food values, for the pot liquor is always served as part of the dish and never thrown down the drain.

A quick-cooking recipe for turnip roots is to dice them, sauté in soy, corn, or olive oil, cover, and simmer until tender, adding a little water or milk if needed. Some people mash them when done, but it is not necessary. Garnish with chives. Our favorite form of turnip is sliced raw in matchstick lengths for what Adelle Davis calls finger salads. We combine them with raw carrot, raw cauliflower, celery, and raw pepper, and serve them with a garlic-and-lentil dip and a mild cheese dip.

watercress

nasturtium officinale

This crisp, refreshing green belongs to the same family as the turnip, but is to be treated entirely differently. We had in our area for many years a good priest who went around planting watercress in clear cool brooks in various places. I go every spring to one or another wild or escaped planting, admire the extent it has spread during the year, and pick a bagful to take home to keep airtight in the refrigerator. I do uproot a few plants to put in the low birdbath we have under a juniper, where they keep on growing for a month or more. You could put them in pots, and keep the pots in a tray of water. You can also grow your own watercress from seed in pots, adding a big handful of leaf mold to the soil and some sand and ground limestone. You can even take some sprigs and root them in a glass of water. Change the water daily.

watercress . . . The best wild watercress bed I know grows on a stream in a limy area, just above a white, lime-bottomed little pond; the worst I know is in a backwater of a little stream that flows by a big storehouse, where there is a great deal of oil and gunk. A few years ago I would pick good cress there several times each spring, but alas, no more.

Good, clean, fresh watercress on bread and butter makes one of the best sandwiches in the world. This herb also adds tang and flavor to the green drink concoction that organic food enthusiasts make in the blender. In this drink they also put weeds.

weeds

Organic gardeners do not hate weeds as much as other people do. Some we revere as pest repellents; others we relish as substitutes for spinach, including amaranthus or pigweed, *Amaranthus retroflexus;* burdock, *Arctium lappa* (p. 160); cattail, *Typha latifolia;* curly dock, *Rumex crispus;* dandelion, *Taraxacum officinale;* lamb's-quarters, *Chenopodium album;* marsh marigold, *Caltha palustris;* milkweed, *Asclepias syriaca* (p. 168); nettles, *Urtica dioica* (p. 169); very young stalks of poke, *Phytolacca americana* (the roots and old stalks are poisonous); purslane, *Portulaca oleracea* (p. 170); Russian thistle, *Salsola pestifer;* violets, *Viola papilionacea;* watercress, *Nasturtium officinale;* wild lettuce, *Lactuca;* and wild primrose, *Primula;* as well as members of the onion family like wild leek, *Allium moly* or *A. tricoccum;* and wild garlic, *Allium vineale.* A few bulbs are tasty, such as those of spatterdock, *Nuphar advena;* toothwort, *Lathraea squamaria;* wild lily, *Lilium philadelphicum;* and cattail, *Typha latifolia*—and, according to Euell Gibbons, many other bulbs. So are some of the roots. For salads, greens, juices to add to soups and gravies or to biscuits and muffins, juices to make into jelly, and tender leaves or slivers of root to make into tempura, wild plants are useful and exciting. As long as you know what you are doing, they add much to the variety and nutrition of an organic gardener's diet. But do not fool with plants you do not know. Consult the books on edible wild plants listed in the Appendix for guidance.

The nutritional values are sometimes phenomenal. For example, the ascorbic acid value of violet leaves and blossoms is way up to 210 and 150 milligrams per 100 grams of plant. Winter cress is also high, with 163 milligrams in the buds and 152 milligrams in the leaves. The only higher source of vitamin C is the leaves of the wild strawberry. If you eat those, you get 229. Highbush cranberry, a common plant in our area, yields 100, which is very good in comparison to dandelion (30), nettle leaves (83), and day lily buds (43). The weeds with good protein percentages are dandelion buds, day lily buds, but especially nettle leaves, with 6.9 grams per 100 grams of protein. For carotene, the most valuable sources are wild spearmint, violet leaves, catnip leaves, and winter cress—both leaves and buds. Figures like this show that violet leaves will give you in half a cup the equivalent in vitamin C of four oranges. We eat the young leaves and buds every spring and like them very much. To us they seem to taste like a cross between spinach and asparagus. You can also make violet flower jelly and syrup. Both should be kept in the refrigerator.

Wild catnip is also high in both vitamins A and C. It should be gathered in July or whenever the blossoms first appear. It is used mostly for tea, for a mild stimulant and also a nervine.

chapter seven

there are still resisters

In spite of what we have known in the past and in spite of what biochemists and plant physiologists are discovering every day now, there are still people who resist the tenets of organic gardening and farming. In spite of the known practices on farms where chemicals were not and have not been used, and in spite of many words of testimony about the successes these farmers and gardeners have had, there are still pockets of deep-seated reluctance to avowing oneself an organic gardener or farmer. There is even a tendency in certain quarters to be suspicious of those who do.

There seems to be quite an array of reasons for this resistance. One is a typical human combination of ignorance and prejudice. Some who know very little about organic gardening methods, and still less about the physiological reasons behind it, do know about rumors reporting that no one but "a food faddist" practices those methods. (Few do know, however, that exactly such rumors were spread around by chemical fertilizer and economic poison salesmen, and invented by their bosses.)

As a matter of fact, the people called "food faddists" were and are very particular about their food; and today they have several million more people in their fold than they did when they hit the scene twenty or thirty years ago. Back then these particular people objected to DDT. They still object to it, both in their food and in their bodies and as it works its nonbiodegradable way through the food chain, threatening many forms of life. They warned, early, what the results now are known to be: every average American has right this minute twelve parts per million of DDT in his own system. It has been found even in newborn babies.

You can grow your own food and fend off a goodly percentage of that twelve parts per million. Some people have resisted doing this; some could not do it—so preferred not to think of the possible results if they had to exist on foods grown by others and processed and brought to market by methods they knew nothing about. The easiest thing to do is to deny that anything could happen to you, or to say: "No one has yet died from having had twelve parts per million of DDT in his system, so why worry?" The next jump of logic, then, is to aver that since no one has been said to die from DDT, it is harmless.

Few people learned in school that one of the main causes of the decline and fall of the Roman Empire was their continued perennial use of lead mugs to drink out of, lead pipes to bring in their water from the hills, cosmetics with lead in them, and lead in their cooking utensils and plates.

Few people are aware that the largest percentage of funds now being expended for the study of pest control at federal biological research stations is being spent for biological control experiments. Whatever they have heard, many people disregard the biological advances already made and believe the pessimistic statement that if we in the United States do not use the powerful hard pesticides and chemical fertilizers, we shall not be able to feed the world—as we have to. Or so we say.

They do know, if they have read volumes like the yearbook of the United States Department of Agriculture on soils, that the federal and state centers of agricultural advice to farmers are usually focused on methods economically feasible—judging cultivation practices not on the basis of what is best for the crop, the people who will eat it, and the soil itself, but on which dollar return will make it worthwhile to do what. The profitability of nitrogen expenditures on corn are figured down to the last penny, for instance. Obviously there is a certain check—for out-and-out ruination of the soil is economically brainless. Now that we have no more frontier this is very obvious, but for years farmers ruined American soils and moved on to another location when they no longer could survive where they were.

In fact, I think one reason why so many people today are rallying to Earth Day calls and ecology action groups is a deep and persistent wish to make up for past devastation of the land. As one young man recently put it when he was telling of his enthusiasm for country living on an organic farm, "We don't have to go and rape the environment to make $50,000 a year like a good capitalist should."

This movement to make up for old ills by returning to the land what we take from it was started in a very small way—with such people as Louis Bromfield and the Friends *resisters . . .* of the Land who met at his farm in the early forties. His books and his followers popularized his efforts to bring back the old exhausted Ohio farmland where he lived to a new bloom of fertility, by organic methods. Edward H. Faulkner, who wrote *Plowman's Folly,* was another who recovered ruined land. His own family had devastated several farms for one generation after another, and he vowed to atone.

But the majority, feeling no such drive, called the Bromfields and the Faulkners of the period nuts. And those who mixed in talk of nutrition with their talk of organic farming were called crackpots. Objectors put their faith in the fertilizer triad, NPK, and saw nothing to shake a belief that chemistry proved and practice showed that NPK would improve crops. Many preferred not to use compost and the natural sources of fertilizer, as the organic gardeners did.

There are lots of odd little reasons why people resist composting. One is a dislike of having things secondhand: hand-me-down clothes, water used before by someone else (more than 100 million of us nevertheless have to use such water), recently decayed old animal tissue and plant tissue spread on the garden where one's food is growing, when instead you could have a clean-looking white or gray substance, out of a new bag, bought at a store.

Along with this is a dislike of manure. I have watched two women at a garden counter at the dime store. One scooped up a bag of dried cow manure, as it said on the label, and put it right under her arm to take to the cashier. The bag was small, the weight five pounds, the substance mealy and sifted, and the smell practically nonexistent. The other woman looked at the name and recoiled. She picked the bag that said *peat humus,* a name which gave her no offense. Organic gardeners like Faulkner and Sir Albert Howard are glad when the materials they use are not new, are obviously being recycled and returned to the source. One of the phrases they use is that these materials "are not from newly broken-down crystals or commercial sources."

I suppose a clue here is that we are so store-oriented that we have a continual craving to buy new materials and resist what nature provides so easily. Even city people can use their garbage for a garden, not to mention the dumpings from the kitty pan.

You have to remember that the farmer who recently arrived at an affluence which enabled him to buy fertilizer was glad enough to give up shoveling manure. New social and financial gains would be hard to break; a return to manuring would not appeal—even though the curing of manure in compost heaps as advised by organic gardeners and farmers lessens the unattractiveness and smell of raw manure 100 percent.

Moreover, various sources of information that influence farmers keep preaching that commercial fertilizers are the way to get affluent. They also preach that nitrogen is nitrogen no matter where it comes from and no matter in what complex pattern it exists and moves.

Many people have heard of the nitrogen cycle, and quite a few of them do know that some sort of bacteria are involved, especially if they have heard of the nodules on legume plants where the nitrifying bacteria live. Some know that bacteria thrive on humus or organic matter in the soil. Most know that organic matter in the soil is good for its structure and aeration. But many people are caught in the prejudice that it is old-fashioned to believe that plant nutrition comes from humus, for the proponents of chemical fertilizers ever since Justus von Liebig have considered that an outmoded belief. The proponents of organic gardening, however, from the first recoil have kept right on thinking that organic materials are important to the full natural nitrogen cycle.

In many instances their knowledge of chemistry was as small as is the knowledge in some of the minds of those who resist. Unfortunately, most of us never get beyond first-year chemistry, or inorganic chemistry. The mysteries of the organic world and its chemistry and its other interrelationships often remain closed to us. People are all too likely to interpret in terms of lifeless, not living, criteria—because they are simple and clear. It is all so much easier that way.

It is also easier, and of course more satisfying to our love of using the brute-strength way of solving things, to approach a system going wrong by getting up and attacking. When monoculture, farm mechanization, and commercially fertilized crops in monoculture attracted pests to the feast, as it was only natural they would, we invented the attack of hard pesticide poisons. Actually some Swiss scientist found DDT decades ago, but it was neglected until it worked wonders and got a good press during World War II because it did such a good job of killing fleas when poured inside soldiers' shirts and because it killed malaria mosquitoes. Now, of course, we have hybrid corn crops in the Midwest that won't respond and the encephalitis or sleeping sickness mosquito that won't respond.

resisters . . .

There are two ironies. With a typical faith that our flying machines and powerful spraying techniques will get every single mosquito (or other pest), we overlook the fact that a few escape. Those who have survived have turned out to be very rugged individuals capable of propagating huge colonies of resistant offspring. A friend of mine told me that even before World War II ended, the flies in her cow barn, treated by DDT which her husband sent her from a war zone, had become resistant to DDT within fourteen months. The other irony is that after discovering this answer, we make up new ones which lead us back to exactly the same error as the one we made in the first place.

Over and over again you hear the answer that if we don't use pesticides, American farming will collapse. And it has gone so far that even small gardeners have got the feeling that their home gardens could not grow to maturity without pesticides—which is nonsense. There is, however, a large grain of truth regarding large monoculture farms: it may actually by now be too late to do anything to stop the vicious spiral started there. The argument now is: "Pesticides not only determine the essential economic factors of our ability to have certain products; pesticides also are essential to managing the pollution of all types of vermin and disease organisms within our homes, schools, public places, business offices—in fact, every element of our daily living," as was said to a House Agriculture Committee hearing by Clifford G. McIntire, director of the American Farm Bureau Federation's natural resources department. He contended that by now both domestic and foreign consumers are dependent on the capacity of the American farmer "to adopt new methods, adjust to less help on the farm, meet demands for capital, and large demands for quality in both fresh and processed foods," and (though he did not say this) to stick to something they think now is really working for them. He did say: "The capacity to do this has taken America over the threshold to an abundant supply of food of the highest quality and safety in the history of any nation."

Organic gardeners do not exactly believe that one. It is natural enough that lobbyists for farmers and their present methods should. Probably many of them do not even know what happens to their product in processing, adulteration, irradiation, and whatever else is done by heat, rays, and chemicals before it reaches the consumer.

Another odd factor is the appeal of gigantism to Americans. We love big things, big operations, big machines to do them with. Those who like them resist the warnings and pleas of those who do not fall for it. Well, we all fall for it one way or another—organic gardeners, like all the rest, do love to hold up huge cabbages, tomatoes, and squashes grown on deep compost and to have their pictures taken and published.

During World War II the standard for measuring foods was switched from quantity to weight. One result of this has been that the bigger the peas or beans or corn per unit that the farmer can put on the market, the better the financial return for him. Customers no longer can shop purely for the best quality per peck. Now we blow up plants and fruits to big size, even as we blow up bread with air and fluffy, almost lifeless starch and turkeys with stilbesterol and water. Obviously, the older peas are, the more giant-sized they become and the fewer it takes to fill up a can or frozen-food box or one-pound bag at the market. The fact that proteins are lost and carbohydrates increased in older vegetables doesn't bother the processor or greengrocer. He probably knows nothing about it. It hardly bothers many housewives any more because we have been conditioned to like something (almost anything) bright green and have by now just about forgotten that the protein value of good young peas is what we really want for our families. When we do not know these things—and thus do not care—we become unwitting resisters to organic home gardening. Even gourmets who have never tasted truly fresh, tender young peas are resisters without knowing it, for they accept what they are offered as "of highest quality and safety."

Resisters include many commercial growers who want to produce the big vegetables fastest, and who therefore use large quantities of nitrogen fertilizers. In your own kitchen on your own plant shelf, big, tall, frail overnitrogened house plants do no one any harm, nor do they bother anyone's palate. Even if the plant topples over from weak stems, it is only an inconvenience. Yet overnitrated, fast-grown spinach, lettuce, and Swiss chard can distort the food value of plants and, when combined with certain amino acids, can form substances of dubious safety to eat. Nitrates in infant food have been challenged by biologists as dangerous. Their effect, when combined with food preservatives for long shelf life, is suspected of being toxic. But these are vague unknowns to resisters, and they tend to brush such things off.

A sad group of resisters is the generation of hamburger eaters, whose craving for more and more hot dogs, *resisters . . .* hamburgers, white buns, bags of crackers and popcorn, and gallons of imitation, gelatinized dairy ices is never satisfied because they are simply not getting the nutrition their body needs with such a diet of calories, sugars, starches, pops and crackles and puffs. They just keep buying more and more, obviously exploited, but stuck in their old habits. Eventually they do not like fresh fruits and vegetables, unprocessed and without additives. As long as they do not change their eating habits, they are unconscious resisters to organic foods and organic methods.

As one defeatist Bostonian complained: "The old realists want the real thing. Well, we haven't got it, and we can't afford it."

Today, if you can become steward of a piece of land and have 35 cents for a packet of seeds, you can afford dozens of heads of lettuce, real and fresh, or a hundred pounds of winter squash, real and sturdy enough to last well into the next winter.

how to resist the resisters

The best way to get the word around and convince resisters is to grow good vegetables, ask people to visit and look at the organic gardens, and, in fact, turn your gardens into demonstration centers. Let the skeptics see for themselves how successful they can be. And how beautiful sometimes, when decorated here and there with bright nasturtiums, neat white feverfew, blooming mints, and clean, rugged, dark-green parsleys and other herbs.

In our area the gardens for demonstration that I've visited run all the way from the most meticulously kept, neatly mulched gardens through equally neat unmulched, but deeply composted gardens to the unorganized, nature-imitating mixture of many different kinds of plants all growing together lushly and amicably—supporting each other both above ground and underground. One garden was an extensive combination of many companionate and deep-rooted plants and healthful herbs, set in long straight rows widely spaced because the owner likes to cultivate and turn his fertilizers in with his rotary turnplow.

start organic gardening clubs

Another good way to spread the word is to establish organic gardening clubs and encourage people to come and hear about others' experiences. And offer to go to talk to already founded garden clubs. Also join the work of ecology groups, pollution fighters, organizations working for recycling collections and detergent-abatement programs. Look up and join natural or organic food cooperatives. If you decide to grow vegetables, fruits, and nuts for profit, develop markets among the people who work in these groups.

gardening clubs . . .

how to convert the resisters

If you do not grow fruits and nuts yourself, send away for some. Bring out some of your best vegetables and other produce, and get people to taste how good organically grown, untreated fruits, nuts, and vegetables can be. Get them to taste utterly fresh foods. Explain how the nutritional values can be kept from dissipating by certain protections you can give them, and how you don't have to throw them away as Americans so often do—through carelessness, wastefulness, or ignorance.

Write articles for bulletins, papers, magazines, and any others who are willing to publish what you have to say. And give talks to clubs and on the radio to tell the public what organic gardening is. There are still people who have never heard of it. Take photographs and show slides. Explain about organic sprays and companionate planting. Many are eager to learn what these mean as soon as they hear of them.

the Americans who seek new ways

From any of these activities—or just from reading the papers—you will find that there are hundreds of Americans now seeking new ways to feed themselves more adequately and more cheaply than they have been doing; who are eager to find a way of life that is nearer the land and the natural, symbiotic relations they know can be enjoyed with other forms of animal and plant life; and who can see that the child born now (and they are being born at the rate of one every seven seconds) is going to need a lot of healthy food and gallons of clean water from the moment he starts crying on his entry into the world. In fact, it is estimated that he is going to ask for 56 million gallons of water, and that is a lot of water. He is going to ask for 20,000 gallons of gasoline and 10,150 pounds of meat and 9,000 pounds of wheat and 28,000 pounds of milk and cream in addition to all the vegetables, fruits, and nuts he can eat and perhaps grow.

From the way things are now going, and the way the ranks of new converts are swelling, it looks as though it might be possible to persuade the resisters before it is too late. And perhaps, in the process, it will be possible to get some of them to put their minds on that 56 million gallons per person demand for water that may turn out to raise a good many more problems than the chemical fertilizer, herbicide, and pesticide threats ever did.

I know there are still many people in this country who not only have never heard of organic gardening but who have scarcely heard of soil erosion either, and have never read that poisonous hard pesticides have inadvertently developed resistant strains of pests now hardened and immune to pesticide sprays. Most of them, however, have heard of food additives and nitrates in baby food and mercury in fish. And I believe that this new awareness means that every one of these people is beginning to have an urge to be an organic gardener—to grow his own vegetables, fruits, berries, and nuts in an environment where creatures and plants can thrive and live in harmony with one another.

appendix and index

appendix

bibliography and directory of addresses

If you are all set now to become an organic gardener—or to expand your present practices—you will probably need to know how to get hold of some of the necessities beyond what you can find in your neighborhood. Perhaps you can even persuade your local store (or start a new store yourself) to provide natural fertilizers, organically grown seeds and plants to set out. Perhaps you want to get more advice and extra supplies from sources supplying organically grown foods. The following lists are compiled to help you do any of those things.

The greatest aids in making these lists have come from the publications of the Rodale Press in Emmaus, Pennsylvania. Any organic gardener should be grateful to the tireless people at this company who keep abreast of developments in the field, sources for help for others, and a vast amount of scientific information coming out every month. We also are grateful that they provide addresses of others interested in the same things we are. Their magazine *Organic Gardening and Farming* is an entertaining and helpful source of hints on every imaginable aspect of organic gardening, with a classified ad section which is a directory in itself.

Other aids were the appendix of Beatrice Trum Hunter's *Gardening Without Poisons* and the publications of the Natural Food and Farming and Bio-Dynamics organizations in this country, and of the Soil Association of Haughley, Suffolk, England.

basic botany

Baker, H. G. PLANTS AND CIVILIZATION. Belmont, Calif.: Wadsworth, 1970.

Bold, H. C. THE PLANT KINGDOM. Englewood Cliffs, N.J.: Prentice-Hall, 1970.

basic botany . . .

Fogg, John M., Jr. WEEDS OF LAWN AND GARDEN. Philadelphia: University of Pennsylvania Press, 1945.

Galston, A. W. THE LIFE OF THE GREEN PLANT. 2d ed. Englewood Cliffs, N.J.: Prentice-Hall, 1964.

Klein, Richard M. and Deana T. RESEARCH METHODS IN PLANT SCIENCE. Garden City, N.Y.: Natural History Press, 1970.

Leopold, A. Carl. PLANT GROWTH AND DEVELOPMENT. New York: McGraw-Hill, 1964.

Loeb, Jacques. REGENERATION. New York: McGraw-Hill, 1924. Specialized studies of plant growth factors.

Machlis, Leonard, and Torrey, John G. PLANTS IN ACTION. San Francisco: W. H. Freeman, 1959.

Martin, W. Coda. A MATTER OF LIFE. New York: Devin-Adair, 1964.

Peattie, Donald Culross. A NATURAL HISTORY OF TREES. Boston: Houghton Mifflin, 1950.

Platt, Rutherford. THIS GREEN WORLD. New York: Dodd, Mead, 1963.

Sinnott, Edmund W. (ed.). BOTANY: PRINCIPLES AND PROBLEMS. New York: McGraw-Hill, 1957.

Steward, F. C. PLANTS AT WORK. Reading, Mass.: Addison-Wesley, 1964.

Sussman, A. S. BIOLOGY THROUGH MICROBES. Ann Arbor: University of Michigan Press, 1961.

Torrey, J. G. DEVELOPMENT IN FLOWERING PLANTS. New York: Macmillan, 1967.

Vallin, Jean. THE PLANT WORLD. New York: Sterling, 1967.

Weizäcker, C. F., von. THE HISTORY OF NATURE. Illinois: University of Chicago Press, 1949.

Wilson, Carl L. BOTANY. New York: Dryden, 1952.

birdhouse kits and plans

Audubon Workshop, Glenview, Ill. 60025.

birds

Baker et al. THE AUDUBON GUIDE TO ATTRACTING BIRDS. New York: Doubleday, 1965.

Halle, Louis J., Jr. BIRDS AGAINST MEN. New York: Viking, 1938.

Hausman, Leon Augustus. THE ILLUSTRATED ENCYCLOPEDIA OF AMERICAN BIRDS. Garden City, N.Y.: Garden City Publishing Co., 1947.

Lemmon, Robert S. HOW TO ATTRACT BIRDS. American Garden Guild, 1962.

McElroy, Thomas P. HANDBOOK OF ATTRACTING BIRDS. New York: Knopf, 1960.

Martin, Alexander C., Zim, Herbert S., and Nelson, Arnold L. AMERICAN WILDLIFE AND PLANTS. New York: McGraw-Hill, 1951.

birds . . .

Peterson, Roger Tory. FIELD GUIDE TO THE BIRDS. Boston: Houghton Mifflin, 1947.

——— and Fisher, James. WILD AMERICA. Boston: Houghton Mifflin, 1955.

Wright, Mabel Osgood. BIRDCRAFT. New York: Macmillan, 1936.

catalogs

(see also *Seeds and Nursery Stock*)

THE WHOLE EARTH CATALOG, 558 Santa Cruz Ave., Menlo Park, Calif. 94025.

THE WINEMAKERS CATALOG. E. S. Kraus, Box 451-C, Nevada, Missouri 64772.

composting

COMPOST SCIENCE, ed. by Jerome Goldstein (monthly), Rodale Press, Emmaus, Pa. 18049.

Rodale, J. I. (ed.). THE COMPLETE BOOK OF COMPOSTING. Emmaus, Pa.: Rodale, 1960.

———. MAKE COMPOST IN 14 DAYS. Emmaus, Pa.: Rodale, 1968.

cookbooks and nutrition

Abehsera, Michel. COOKING FOR LIFE: A GUIDE FOR THE WELL-BEING OF MAN. Binghamton, N.Y.: Swan House, 1970.

———. ZEN MACROBIOTIC COOKING. Binghamton, N.Y.: Swan House, 1969.

Beck, Bodog F., and Smedley, Dorée. HONEY AND YOUR HEALTH. New York: Bantam, 1944.

Chen, Philip. SOYBEANS FOR HEALTH. Emmaus, Pa.: Rodale, 1969.

Davis, Adelle. LET'S COOK IT RIGHT. New York: Harcourt, Brace, 1970.

———. LET'S EAT RIGHT TO KEEP FIT. New York: Harcourt, Brace, 1954.

Farmer, Fannie Merritt. THE BOSTON COOKING SCHOOL COOK BOOK. Boston: Little, Brown, 1937.

Hittleman, Richard. YOGA NATURAL FOODS COOKBOOK. New York: Bantam, 1970.

Hunter, Beatrice Trum. THE NATURAL FOODS COOKBOOK. New York: Simon and Schuster, 1961.

Hunter, Kathleen. HEALTH FOODS AND HERBS. New York: Wm. Collins, 1970.

Jones, Dorothea Van Gundy. THE SOYBEAN COOKBOOK. New York: Arc, 1971.

Kagawa, Aya. JAPANESE COOKBOOK. Japan: Japanese Travel Bureau, 1963.

Omarr, Sydney, and Roy, Mike. COOKING WITH ASTROLOGY. New York: Signet Mystic, 1969.

cookbooks . . . Ohsawa, Georges. THE PHILOSOPHY OF ORIENTAL MEDICINE: Vol. II: THE BOOK OF JUDGMENT. Paris, 1964.

Rawlings, Marjorie Kinnan. CROSS CREEK COOKERY. New York: Scribner's, 1942.

Rodale, J. I., and Staff. THE COMPLETE BOOK OF VITAMINS. Emmaus, Pa.: Rodale, 1966.

THE TUESDAY SOUL FOOD COOKBOOK. New York: Bantam, 1969.

Toklas, Alice B. THE ALICE B. TOKLAS COOK BOOK. New York: Doubleday, 1960.

dormant oil spray

Scalecide (trade name), B. G. Pratt Co., 206 21st Ave., Paterson, N.J. 07503.

earthworms

Andrew Peoples, R.D. 1, Lansdale, Pa. 19445.

Brazos Worm Farms, Rt. 9, Waco, Tex. 76705.

Darwin, Charles. DARWIN ON HUMUS AND THE EARTHWORM. Mystic, Conn.: Verry, 1966.

ecology

Billings, W. D. PLANTS AND THE ECOSYSTEM. Belmont, Calif.: Wadsworth, 1969.

Borland, Hal. BEYOND YOUR DOORSTEP: A HANDBOOK TO THE COUNTRY. New York: Knopf, 1962.

Buschbaum, Ralph and Mildred. BASIC ECOLOGY. Pittsburgh: Boxwood, 1957.

Carson, Rachel. SILENT SPRING. Greenwich, Conn.: Fawcett, 1962.

Dansereau, Pierre (ed.). CHALLENGE FOR SURVIVAL: LAND, AIR AND WATER FOR MAN IN MEGALOPOLIS. New York: Columbia University Press, 1970.

DeBell, Garrett. THE ENVIRONMENTAL HANDBOOK. Friends of the Earth; New York: Ballantine, 1970.

Dubos, René. SO HUMAN AN ANIMAL. New York: Scribner's, 1970.

Ehrlich, Paul R. POPULATION BOMB. New York: Ballantine, 1968.

Farb, Peter. THE LIVING EARTH. New York: Harper & Row, 1969.

Graham, Frank, Jr. SINCE SILENT SPRING. Boston: Houghton Mifflin, 1970.

Kormondy, Edward J. CONCEPTS OF ECOLOGY. Englewood Cliffs, N.J.: Prentice-Hall, 1969.

Kourennoff, Paul M. RUSSIAN FOLK MEDICINE. Trans. by George St. George. New York: Pyramid, 1971.

Leopold, Aldo W. SAND COUNTY ALMANAC. New York: Oxford University Press, 1968.

ecology . . .

Lord, Russell. CARE OF THE EARTH. New York: New American Library, 1969.

______. FOREVER THE LAND. New York: Harper, 1950.

Ottinger, Betty Ann. WHAT EVERY WOMAN SHOULD KNOW—AND DO—ABOUT POLLUTION. The EP Press, 1970.

Peters, Harold S. "The Ecology of Pollution," NATURAL FOOD AND FARMING, Vol. 13, No. 7 (Dec., 1966).

A PLACE TO LIVE: THE YEARBOOK OF AGRICULTURE. Washington, D.C.: U.S. Dept. of Agriculture, 1963.

Platt, Rutherford. THE GREAT AMERICAN FOREST. Englewood Cliffs, N.J.: Prentice-Hall, 1965.

______ and Albright, Horace. ADVENTURES IN THE WILDERNESS. New York: Harper, 1963.

Rudd, Robert L. PESTICIDES AND THE LIVING LANDSCAPE. Madison: University of Wisconsin Press, 1964.

Sears, Paul B. DESERTS ON THE MARCH. Norman: University of Oklahoma Press, 1959.

Sierra Club. ECOTACTICS. San Francisco: Sierra Book Club, 1970.

Shepard, Paul, and McKinley, D. MAN IN THE LANDSCAPE. New York: Knopf, 1969.

______. THE SUBVERSIVE SCIENCE. Boston: Houghton Mifflin, 1970.

Storer, J. H. THE WEB OF LIFE. New York: New American Library, 1966.

Thoreau, Henry David. WALDEN. 1864.

ecology centers

Susan "Rabbit" Goody, Boston Area Ecology Action Center, 925 Massachusetts Ave., Cambridge, Mass. 02139.

Stuart Leiderman, Environmental Response, Box 1124, Washington University, St. Louis, Missouri 63105.

Bob Nelson, Midwest Environmental Education and Research Association, 1051 McKnight Road, St. Paul, Minn. 55119.

Pat Shaylor, Ecology Center, 2179 Alliston Way, Berkeley, Calif. 94074.

fish sources

Fish Emulsions, Acme Peat Products, Ltd., 687 North 7 Road, Rt. 2, Richmond, British Columbia, Canada. Liquid whale plant food, bone.

Stapco, Standard Products Company, Inc., White Stone, Virginia.

Alaska Fertilizer Co., 84 Seneca St., Seattle, Washington. 98101.

founders of the organic gardening movement

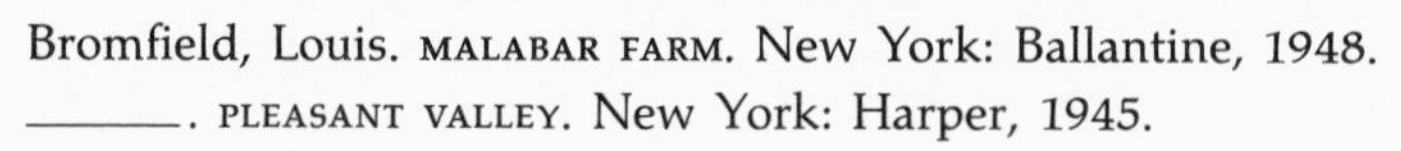

Bromfield, Louis. MALABAR FARM. New York: Ballantine, 1948.

———. PLEASANT VALLEY. New York: Harper, 1945.

Faulkner, Edward H. PLOWMAN'S FOLLY. Norman: University of Oklahoma Press, 1943.

Howard, Sir Albert. AN AGRICULTURAL TESTAMENT. London: Oxford University Press, 1940.

———. THE SOIL AND HEALTH. New York: Devin-Adair, 1947.

Nearing, Helen and Scott. LIVING THE GOOD LIFE. New York: Schocken, 1970.

Pfeiffer, Ehrenfried. BIO-DYNAMIC FARMING AND GARDENING. New York: Anthroposophic Press, 1943.

———. THE EARTH'S FACE AND HUMAN DESTINY. Emmaus, Pa.: Rodale, 1947.

Rodale, J. I. THE ORGANIC FRONT. Emmaus, Pa.: Rodale, 1949.

———. PAY DIRT. Emmaus, Pa.: Rodale, 1971.

fruit

Lee Anderson's Covalda Date Company, 51–392 Highway 36, Box 908, Coachella, Calif. 92236.

L. P. DeWolf, Crescent City, Fla. 32012. Tree-run grade citrus fruit, "organi-culturist."

Golden Acres Orchard, Rt. 2, Box 70, Fort Royal, Va. 22630.

Halbleib Orchards, McNabb, Ill. 61335.

Russell Citrus Groves, Rt. 3, Box 57, Mission, Tex. 87572.

Tex-Organic Fruit Co., Box 147, McAllen, Tex. 78501.

gardening guides

THE AMERICAN HOME GARDEN BOOK AND PLANT ENCYCLOPEDIA. New York: M. Evans, 1963.

Bruning, Walter F. MINIMUM MAINTENANCE GARDENING HANDBOOK. New York: Harper & Row, 1970.

Carleton, R. Milton. THE SMALL GARDEN BOOK. New York: Macmillan, 1971.

Cocannouer, Joseph. FARMING WITH NATURE. Norman: University of Oklahoma Press, 1954.

———. WEEDS: GARDENING WITH THE SOIL. New York: Devin-Adair, 1954.

Cosper, Lloyd C., and Logan, Harry B. HOW TO GROW VEGETABLES. New York: Duell, Sloan and Pearce, 1951.

Cruso, Thalassa. MAKING THINGS GROW. New York: Knopf, 1969.

Darlington, Jeanie. GROW YOUR OWN: AN INTRODUCTION TO ORGANIC GARDENING. Berkeley: The Book People, 1970.

Duncan, Frances. THE JOYOUS ART OF GARDENING. New York: Scribner's, 1917.

gardening guides . . .

Gillespie, Janet. PEACOCK MANURE AND MARIGOLDS. New York: Viking, 1964.

Gregg, Richard. COMPANION PLANTS AND HERBS. Stroudsburg, Pa.: Bio-Dynamic Farming and Gardening Association.

Hiscox, Gardner D. (ed.). HENLEY'S TWENTIETH CENTURY BOOK OF TEN THOUSAND FORMULAS, PROCESSES, AND TRADE SECRETS. New York: Books, 1970.

Hunter, Beatrice Trum. GARDENING WITHOUT POISONS. New York: Berkley, 1971.

Kenfield, Warren G. THE WILD GARDENER IN THE WILD LANDSCAPE. New York: Hafner, 1970.

Leighton, Ann. EARLY AMERICAN GARDENS. Boston: Houghton Mifflin, 1969.

Northen, Henry T. and Rebecca T. THE SECRET OF THE GREEN THUMB. New York: Ronald, 1954.

Ogden, Samuel R. HOW TO GROW FOOD FOR YOUR FAMILY. New York: Barnes, 1942.

_______. STEP-BY-STEP TO ORGANIC VEGETABLE GARDENING. Emmaus, Pa.: Rodale, 1971.

Pellegrini, Angelo M. THE FOOD-LOVER'S GARDEN. New York: Knopf, 1970.

Philbrick, John and Helen. GARDENING FOR HEALTH AND NUTRITION. Free Deed Books, 1963.

Rodale, J. I. (ed.). BEST IDEAS FOR ORGANIC VEGETABLE GARDENING. Emmaus, Pa.: Rodale, 1970.

_______. THE ENCYCLOPEDIA OF ORGANIC GARDENING. Emmaus, Pa.: Rodale, 1959.

_______. THE ORGANIC WAY TO PLANT PROTECTION. Emmaus, Pa.: Rodale, 1966.

Rodale, Robert. THE BASIC BOOK OF ORGANIC GARDENING. New York: Ballantine, 1971.

Sara, Dorothy. THE NEW AMERICAN GARDEN BOOK. New York: Books, 1961.

Seymour, E. L. D. THE NEW GARDEN ENCYCLOPEDIA. New York: Wm. H. Wise, 1936.

Tyler, Hamilton. ORGANIC GARDENING WITHOUT POISONS. New York: Van Nostrand Reinhold, 1970.

geese

Midsouth Weeder Geese, Columbus, Missouri. 65201. Geese rented or sold for weeding.

goats

DAIRY GOAT GUIDE, Countryside Publications, 318 Waterloo Rd., Marshall, Wisc. 53559.

DAIRY GOAT JOURNAL, P.O. 836, Columbia, Missouri 65201.

greenhouses

Aluminum Greenhouses, Inc., 14615 Lorain Ave., Cleveland, Ohio 44111.

Lord and Burnham, Irvington, N.Y. 10533.

J. A. Nearing, 10788 Tucker St., Box 346, Beltsville, Md. 20705.

Redfern Prefab Greenhouses, Dept. H., 55 Mt. Hermon Rd., Scotts Valley, Calif. 95060.

Reimuller Fiberglas Greenhouses, Box 5276, Riverside, Calif. 92507.

Turner Greenhouses, P.O. Box 1260, Goldsboro, N.C. 27530.

herbs

Caprilands Herb Farm, Silver St., Coventry, Conn. 06238.

Casa Yerba, Box 176, Tustin, Calif. 92680.

Coon, Nelson. USING PLANTS FOR HEALING. New York: Hearthside, 1963.

Culpeper, Nicholas. COMPLETE HERBAL. Hackensack, N.J.: Wehman, 1960.

Greene Herb Gardens, Greene, R.I. 02827.

Harvest Health, Inc., 1944 Eastern Ave. S.E., Grand Rapids, Mich.

THE HERB GROWER (quarterly), Falls Village, Conn. 06031.

Kerbel Pharmacy, 1473 Bedford Ave., Brooklyn, N.Y.

Ledoar Nurseries, 7206 Belvedere Rd., W. Palm Beach, Fla. 33406.

Le Jardin du Gourmet, Box 119-8, Ramsey, N.J. 07446. Send for their shallots and leeks for fall planting.

Meadowbrook Herb Garden, Wyoming, R.I. 02898.

Merit Herb Co., South Chicago Station, Chicago, Ill. 60617.

Merry Gardens, Camden, Maine 04843.

Nature's Herb Co., 281 Ellis St., San Francisco, Calif. 94102.

Nichols Garden Nursery, 1190 N. Pacific Hwy., Albany, Ore. 97321.

North Central Herbal Comfrey Producers, Box 195J, Glidden, Wisc. 54527.

Pine Hills Herb Farms, Box 144, Roswell, Ga. 30075.

Simmons, Adelma Grenier. HERB GARDENING IN FIVE SEASONS. New York: Van Nostrand, 1964.

Sunnybrook Farms Nursery, 9448 Mayfield Rd., Chesterland, Ore. 44026.

Webster, Helen Noyes. HERBS, HOW TO GROW THEM AND HOW TO USE THEM. Newton, Mass.: Branford, 1942.

insect controls

(see also *Ladybugs; Praying Mantises*)

Agrilite Systems, Inc., 404 Barringer Building, Columbia, S.C.

Aquacide, IMS Corp., Albuquerque, N.M.

Biotrol (trade name for *Bacillus thüringiensis*), Thompson-Hayward Chemical Co., Box 2382, Kansas City, Kans. 66110.

insect controls . . .

Brooklyn Botanical Garden, "Handbook on Biological Control of Plant Pests," Brooklyn, N.Y.

Catskill Imports, Box H, South Fallsburg, N.Y. 12779.

Doom (trade name for milky spore disease), Fairfax Biological Laboratory, Clinton Corners, N.Y. 12514.

D-Vac Co., Box 2095, Riverside, Calif. 92506.

"How to Kill Insects the Non-Toxic Way . . . with Pyrethrins." Pyrethrum Information Center, Room 423, 744 Broad Street, Newark, N.J. 07102.

Hunter, Beatrice Trum. GARDENING WITHOUT POISONS. New York: Berkley, 1971.

Old Southern Giant Bullfrog Farm, 5607 Booker Rd., Evansville, Ind. 44712.

Philbrick, John and Helen. THE BUG BOOK. Box 96, Wilkinsonville, Mass. 01590.

The Tanglefoot Company, 314 Straight Ave., S.W., Grand Rapids, Mich. 49500.

Thuricide (trade name for *Bacillus thüringiensis*), International Mineral and Chemical Corp., Crop Aid Products, Dept. 5401, Old Orchard Rd., Skokie, Ill. 60076.

Trik-O (trade name for trichogramma wasps), Gothard, Inc., Box 370, Canutillo, Tex. 79835. Wasps for use in flower, vegetable gardens, fruit and nut trees. Control the apple codling moth.

Vitova Insectary, Inc., Box 475, Rialto, Calif. 92376. Lacewings and trichogramma wasps.

kitchen aids

Barth's, Valley Stream, N.Y. 11580. Home flour mills.

Turan, Inc., 5529 Morris St., Philadelphia, Pa. 19144. Corn meal grinder. Also available through Nichols Garden Nursery, 1190 North Pacific Hwy., Albany, Ore. 97321.

Fairbanks Co., 1275 South St. Paul St., Denver, Colo. 80210. Gadget to recover shredded garbage from disposal unit in the sink.

W. R. Laboratories, Box 364, Dept. NF, Lodi, Calif. Juicer-extractor.

Log-cabin Sprouter, 21 Century Foods, 801 South Bloomington, Streator, Ill. 61364. Seed-sprouter.

Westchester Health Shop, Mrs. Ruth Rieger, 367 Elwood Ave., Hawthorne, N.Y. 10532. Yogurt makers.

ladybugs

Bio-Control Co., Rt. 2, Box 2397, Auburn, Calif. 95603.

L. E. Schnoor, P.O. Box 114, Rough and Ready, Calif. 95975.

Lakeland Nurseries, Hanover, Pa. 17331.

World Garden Products, 2 First St., E., Norwalk, Conn. 06855.

magazines

AMERICAN FORESTS. The American Forestry Association, Washington, D.C.

BULLETIN OF THE GARDEN CLUB OF AMERICA, 598 Madison Ave., N.Y.

THE HOMESTEADER, Oxford, N.Y. 13839. Articles concerning organic gardening, the simple life, handicrafts.

THE MOTHER EARTH NEWS, P.O. Box 38, Madison, Ohio 44057.

THE NATIONAL GARDENER, National Council of State Garden Clubs, St. Louis, Missouri. 63110.

ORGANIC GARDENING AND FARMING, Emmaus, Pa. 18049.

natural foods

Deer Valley Farm Natural Foods, Deer Valley Bakery, R.D. 1, Guilford, N.Y. 13780. Natural foods, organically home-baked goods.

Desert Herb Tea Company, 736 Darling St., Ogden, Utah. 84403.

Erewhon Trading Co., Inc., 342 Newbury St., Boston, Mass. 02115.

FOOD FACTS BOOK, Heritage, 205 N. Howard, Woodstown, N.J. 08098.

Goldstein, Jerome, and Goldman, M. C. GUIDE TO ORGANIC FOODS SHOPPING AND ORGANIC LIVING. Emmaus, Pa.: Rodale, 1970. Lists sources for organic foods in 46 states and Canada.

The Good Earth, 1336 First Ave., New York, N.Y. 10011.

Organic Food Center, 557 Bedford St., Whitman, Mass. 02382.

Organic Merchants, 1326 Ninth Ave., San Francisco, Calif. 94122.

Rorty, James, and Norman, N. Philip. BIO-ORGANICS: YOUR FOOD AND YOUR HEALTH. New York: Lancer, 1956.

Snow Hill Farm, R.D. 4, Coatesville, Pa. 19320. Offers a variety of meat products; animals are raised on an organic diet.

Vrest Orton, Vermont Country Store, Weston, Vt. 05161.

Walnut Acres Mill & Store, Penns Creek, Pa. 17862.

Wolfe's Neck Farm, Freeport, Maine 04032. Beef.

netting

Animal Repellents, Box 168, Griffin, Ga. 30223.

Apex Mills, Inc., 49 W. 37th St., New York, N.Y. 10018.

Frank Coviello, 1300 83d St., N. Bergen, N.J. 17047.

not to be missed

Bay Laurel, Alicia. LIVING ON THE EARTH. New York: Vintage, 1971.

THE HOMESTEADER'S BIBLIOGRAPHY, by Elsie Evelsizer, Winterthur Goat Farm, Rt. 2, Forrest, Ill. 61741.

THE EARTH THEATRE, headed by Corinna Harmon (R.D. 1, Pawlet, Vt. 05761) and Glenn Munson; mime, dance, and comedy in operation on problems of pollution, overpopulation, and ecology.

nutrition

See *Cookbooks and Nutrition.*

organic fertilizers and soil conditioners

Alsmith Ind., Box 275, Seaford, N.Y. 11783. Greensand.

F. S. Brisbois, Fonda, Iowa 50540. Kelp.

Brookside Nurseries, Darien, Conn. 06820. Rock silt, composting materials, and a host of soil aids.

Joe S. Francis, Blenders, Inc., Lithonia, Ga. 30058. Organic fertilizers, Hybro-tite, Re-vita, and poultry compost.

Grobecs, Ashvitte, Minn. 55036. Kelp.

Hocking Granite Industries, Clark Island, St. George, Maine. Rock phosphate and granite dust, Vita-mite.

Kaylonite Corp., Dunkirk, Maryland. Greensand.

New Life Soil Conditioner, Box 241, Lewiston, Mont. 59457.

Odlin Organics, Lakeshore Drive, W. Brookfield, Mass. 01585. Mer-Made plant food, liquid fish fertilizer concentrate, natural and organic fertilizers, soil conditioners, minerals, power equipment.

Q. R. Herbal Starter, Bostock, Franconia, N.H. 03580.

The Garden Mart, 5108 Bissonnet St., Bellaire, Tex. Fish and seaweed for root and leaf feeding.

The Pfeiffer Foundation, Inc., Threefold Farm, Spring Valley, N.Y. 10977. B.D. compost starter.

Wonder Life Co., 3824 Douglas Ave., Des Moines, Iowa 50310. Livestock feeds, soil builders.

Zook and Ranck, Inc., Rt. 1, Gap, Pa. 17527.

organic sprays

Peter A. Escher, Threefold Farms, Spring Valley, N.Y. 10977. B-DD and 3-D tree spray.

Gregg, Evelyn. BIO-DYNAMIC SPRAYS. Stroudsburg, Pa.: Bio-Dynamic Farming and Gardening Association.

Hopkins Agricultural Chemical Co., Box 584, Madison, Wisc. 53701.

Natural Development Company, Bainbridge, Pa. 17502. Tri-Evcel DS.

Rotenone, pyrethrum, sabadilla, and ryania are available at most garden supply stores and seed firms.

praying mantises

Eastern Biological Control Co., Rt. 5, Box 379, Jackson, N.J. 08527.
Gothard, Inc., Box 332, Canutillo, Tex. 79835.
Robert Robbins, 424 N. Courtland, East Stroudsburg, Pa. 18301.

rabbits

THE AMERICAN RABBIT JOURNAL, Indianapolis, Ind. White's Rabbitry, Mt. Vernon, Ohio 43050.

seeds and nursery stock

W. Atlee Burpee Co., 427 Burpee Bldg., Philadelphia, Pa. 19132.
Farmer Seed and Nursery Co., Faribault, Minn. 55021.
Gamiel Story, Box 263, Lebanon, Tenn. All shades of sunflowers.
Joseph Harris Co., Moreton Farm, Rochester, N.Y. 14624.
Kelly Brothers Nurseries, Inc., Dansville, N.Y. 14437.
Lakeland Nurseries Sales, Hanover, Pa. 17331.
Orol Ledden and Sons, Center Ave., Sewell, N.J. 08080. Organic seeds.
Henry Leuthardt Nurseries, Inc., Long Island, N.Y.
J. E. Miller Nurseries, Inc., Canandaigua, N.Y. 14424.
Natural Development Co., Bainbridge, Pa. 17052. Edible seeds of soybean, millet, barley, oats, buckwheat, sunflower; stocks Fertrell, an organic fertilizer.
Nichols Garden Nursery, 1190 North Pacific Highway, Albany, Ore. 97321.
George W. Park Seed Co., Greenwood, S.C. 29646.
Holmes Quisenbery, Greenbrier Drive, Athens, Ohio. 45701. Mexican tomatoes.
R. H. Shumway Seedsman, P.O. Box 777, Rockford, Ill. 61101.
Stark Brothers' Nurseries and Orchards, Louisiana, Missouri 63353.
Stokes Seeds, Inc., Box 15, Ellicott St. Station, Buffalo, N.Y. 17205.
Thompson and Morgan, Ipswich, Ltd., Ipswich, England.
Walnut Acres, Penns Creek, Pa. 17862.
Wayside Gardens, Mentor, Ohio 44060.
White Flower Farm, Litchfield, Conn. 06759.

soil

Balfour, Lady Eve. THE LIVING SOIL. London: Faber and Faber, 1943.
Buckman, Harry O., and Brady, Nyle C. THE NATURE AND PROPERTIES OF SOILS. New York: Macmillan, 1969.
Donahue, Roy L. SOILS: AN INTRODUCTION TO SOILS AND PLANT GROWTH. The Ford Foundation; New Delhi, India: Prentice-Hall, 1965.
Forman, Jonathan. SOIL, FOOD AND HEALTH. Atlanta, Tex.: Natural Food Associates, 1961.

Garrett, S. D. SOIL FUNGI AND SOIL FERTILITY. Elmsport, N.Y.: Pergamon, 1969.

soil . . .

Howard, Sir Albert. THE SOIL AND HEALTH: A STUDY OF ORGANIC AGRICULTURE. New York: Devin-Adair, 1947.

Ortloff, H. Stewart, and Raymore, Henry. A BOOK ABOUT SOILS FOR THE HOME GARDENER. New York: Barrows, 1962.

SOILS AND MEN: YEARBOOK OF AGRICULTURE. Washington, D.C.: USDA, 1938.

SOIL: THE YEARBOOK OF AGRICULTURE. Washington, D.C.: USDA, 1957.

Waksman, Selman A., and Starkey, Robert L. THE SOIL AND THE MICROBE. New York: John Wiley, 1931.

Worthen, Edmund L. FARM SOILS: THEIR MANAGEMENT AND FERTILIZATION. New York: John Wiley, 1948.

soil conditioners

See *Organic Fertilizers and Soil Conditioners.*

soil test kits

Indiana Botanic Gardens, Hammond, Ind. 46325.

George W. Park Seed Co., Greenwood, S.C. 29646.

Sudbury Laboratory, Inc., Sudbury, Mass. 01776.

The House Plant Corner, Box 810, Oxford, Md. 21654.

sprays

See *Organic Sprays; Dormant Oil Spray.*

tankage

Brookside Nurseries, Darien, Conn. 06820. Feather, hair, and hide tankage.

tools

H. P. Co., Box 391-E, Wicasset, Maine 04578. Magic Weeder.

Kemp Shredder Co., Box 975, Erie, Pa. 16512. Hammermill shredder-chopper.

Troy-built Roto-Tillers, Troy, N.Y. New machine combines tilling and chopping.

M. A. Johnson, Rt. 5, Box 447, Taylorsville, N.C. 28681. Single-blade shredder-chopper, lightweight, attaches to mower.

Gravely, Gravely Lane, Clemons, N.C. 27012. Rotary turnplows, tractors.

trace elements

Happy Acres, Box 711, Somerset, Ky. 42501.

traps Havahart, 148 Water St., Ossining, N.Y. 10562.
Johnson's, Box 13, Waverly, Ky. 42462.

wastes (municipal composting reclaimed) Altoona FAM, Rm. 468, Altoona Trust Bldg., Altoona, Pa. 16603.
City of Schenectady, N.Y. Orgro, sludge.
Sewerage Committee, Milwaukee, Wisc. Milorganite, sludge.

wild plants

Fernald, Merritt L., and Kinsey, A. C. EDIBLE WILD PLANTS OF EASTERN NORTH AMERICA. New York: Harper & Row, 1958.

Gibbons, Euell. STALKING THE HEALTHFUL HERBS. New York: David McKay, 1966.

———. STALKING THE WILD ASPARAGUS. New York: David McKay, 1962.

Hatfield, Audrey Wynne. PLEASURES OF WILD PLANTS. London: Museum Press, 1966.

MacLeod, Dawn. A BOOK OF HERBS. London: Gerald Duckworth, 1968.

Taylor, Kathryn S. A TRAVELER'S GUIDE TO ROADSIDE WILD FLOWERS, SHRUBS AND TREES OF THE UNITED STATES. New York: Farrar, Straus, 1949.

Taylor, Norman, WILD FLOWER GARDENING. Princeton, N.J.: D. Van Nostrand, 1955.

index

A Note about the Author

Catharine Osgood Foster was born in Newton Highlands, Massachusetts, and was graduated from Mount Holyoke College, but became a Vermonter by adoption when she started teaching literature at Bennington College and later married a native, Thomas H. Foster, ornithologist and critic. She has been actively involved in conservation work in the state of Vermont for many years and since her retirement from teaching she has written a weekly column for the Bennington *Banner.* The Fosters live near Old Bennington, where they practice what they preach surrounded by trees and shrubs, a garden and a yard with living creatures who help maintain the good health and harmony of their land.

A Note about the Type

The text of this book was set in Elegante, the film version of Palatino, a type face designed by the noted German typographer Hermann Zapf. Named after Giovambattista Palatino, a writing master of Renaissance Italy, Palatino was the first of Zapf's type faces to be introduced to America. The first designs for the face were made in 1948, and the fonts for the complete face were issued between 1950 and 1952. Like all Zapf-designed type faces, Palatino is beautifully balanced and exceedingly readable.

The book was composed by Graphic Services, Inc., York, Pennsylvania; and designed by Emil Antonucci.

HB4D